William Townsend Porter

The growth of St. Louis children

William Townsend Porter
The growth of St. Louis children
ISBN/EAN: 9783337821432
Printed in Europe, USA, Canada, Australia, Japan
Cover: Foto ©ninafisch / pixelio.de

More available books at **www.hansebooks.com**

Transactions of The Academy of Science of St. Louis.

VOL. VI.   No. 12.

THE GROWTH OF ST. LOUIS CHILDREN.

WILLIAM TOWNSEND PORTER.

Issued April 14, 1894.

THE GROWTH OF ST. LOUIS CHILDREN.

INTRODUCTION.

In November, 1891, I applied to the Teachers' Committee of the St. Louis Board of Public Schools for permission to make a series of physical measurements of the St. Louis school children. The ends in view were the study of the phenomena of growth, the making of physical standards for each age in the period of school life, and the adjustment of school tasks to the pupil's strength. On the recommendation of Mr. Long, superintendent of the Public Schools, Dr. Hickman and Mr. Walter F. McEntire, at that time chairman of the committee, it was resolved to lay before the Board a statement of the purpose of the measurements and to advise that the permission to make them be granted. This favorable report, for which thanks are due the gentlemen just named, caused the Board to authorize the measurements at its sitting December 8, 1891.

The measurements were collected by what statisticians know as the generalizing method. In the generalizing method, a great number of children is measured once, and the measurements classified according to age. The mean height of the boys or girls at each age is regarded as the height typical of that age. When these typical heights are arranged in order, they show the increase in the height of the type-child during his period of growth and thus express a law of growth. A similar procedure reveals the growth in weight, girth of chest, or any other physical dimension. It is believed that the values got by the generalizing method are the same as would be obtained if a smaller number of children was measured yearly during the growth period. In either case, the accuracy of the result depends on the number of observations at each age, and a high degree of accuracy requires the making of many thousand measurements.

So large an undertaking demands great labor and a considerable expenditure of money. The labor was in the present instance shared by many hands. Most of the measurements were made by the teachers. The measurements of the head and face were made by undergraduates of the St. Louis Medical College. Other members of the same institution were of the greatest service as "special assistants." Messrs. Taussig, Gooden, Soper, Blair, Gross, Schlossstein, Lemen, Loth, Newcomb and Simmons served in this capacity. To their unwearied and long-continued labors and to the support unselfishly given them by their fellow-students the success of the work is chiefly due.

The expense of the investigation was for apparatus, printing and the hire of clerks, and was borne by gentlemen of public spirit in St. Louis. Some of the apparatus was given or loaned free of charge. The Simmons Hardware Company gave callipers and measuring tapes; the Fairbanks Scale Company loaned scales; Mr. F. W. Humphrey loaned a dozen watches; Dr. John Green and Dr. A. E. Ewing gave cards for testing the acuteness of vision, and the Nixon-Jones Printing Company made special rates for printing.

The cordial support of Chancellor Chaplin, Professors H. S. Pritchett, G. Hambach and E. A. Engler of Washington University, the valuable advice of Dr. John B. Shapleigh, and the interest of the Academy of Science of St. Louis are gratefully acknowledged.

The data collected are necessarily of unequal value. Many of the curves constructed from them are highly satisfactory; others present irregularities to be ascribed to an insufficient number of observations at those points. Some investigators have withheld the curves in which such irregularities occur. They have all been printed here, because the wide-spread use of anthropometrical methods in the Public Schools, so much to be desired, seems at present only possible where the measurements are made by the teachers, and it is necessary to know by the examination of the total material of such investigations, what degree of accuracy can be expected.

CHAPTER I.

THE COLLECTION OF THE MEASUREMENTS.

In collecting anthropometrical measurements on a large scale, a systematic working-plan is evidently of much importance. Many things should be borne in mind in making such a plan. The school routine should be disturbed as little as possible; the directions to the measurers should be comprehensive and clear; the apparatus employed should be simple; the measurements should be made during a few months only and in the same season of the year, for the weight of clothing and even the rate of growth differs at different seasons, and measurements collected partly in one season and partly in another cannot be so suitable for comparison as those collected only in the winter or only in the summer; the order in which the schools are visited should be governed by their situation, so that no time may be spent unnecessarily in going and coming; and much care should be taken to collect sufficient data concerning the social condition and the nationality of parents. In short, the construction of a serviceable working-plan for anthropometrical measurements of great numbers of children is a difficult task, and it is to be regretted that the literature is all but barren of suggestions as to the best method of collecting such material. It is hoped for this reason that a description of the methods used in this investigation may be not without value.

The data collected are mentioned in Form A., one of four "forms" employed in this investigation.

Form A.

MALE.

Please Write with Ink.

1. Observer..
2. School..
3. Date..
4. Name of pupil..
5. Place of birth...
6. Age at nearest birthday..

7. In what country was father born ....
8. In what country was mother born..................................
9. Occupation of father..
10. No. of sisters living........................; dead....................
11. No. of brothers living.......................; dead....................
12. Residence, No..Street.

13. Hair { Black. / Dark Brown. / Light Brown. / Red. / Flaxen. }   14. Eyes { Dark Brown. / Light Brown. / Blue. / Grey. }

15. Height standing...cm.
16. Height sitting...cm.
17. Span of arms..cm.
18. Strength of squeeze, right hand..lbs.
19. Strength of squeeze, left hand...lbs.
20. Girth of chest, forced expiration..cm.
21. Girth of chest, forced inspiration.......................................cm.
22. Weight......................lbs.
23. Acuteness of vision, right eye...
24. Acuteness of vision, left eye..
25. Acuteness of hearing, right ear..................ft..............in
26. Acuteness of hearing, left ear...................ft..............in.
27. Length of head...mm.
28. Width of head..mm.
29. Height of face..mm.
30. Width of face...mm.
31. The height of face from the hair line to the point of chin...........mm.
32. Grade...

Form B was the same as Form A, except that the paper on which it was printed was green instead of white, the color of Form A, and the word "Male" was replaced by the word "Female."

On Jan. 4, 1892, the first school * in the series to be examined began to set down the answers 1, 2, 3, 4, 13, 14, 32, and its pupils were directed to carry home Form C in order that it might be filled by the parents.

---

* Pupils were measured in the following schools: Ames, Blair, Blow, Branch High, Bryan Hill, Carr, Carroll, Carr Lane, Central High, Charless, Chouteau, Clay, Clinton, Crow, Divoll, Douglass, Eliot, Elleardsville, Franklin, Garfield, Hamilton, Hodgen, Humboldt, Irving, Jackson, Jefferson, Laclede, Lafayette, Lincoln, Lowell, Lyon, Madison, Maramec, Mullanphy, O'Fallon, Peabody, Penrose, Pestallozzi, Polytechnic, Pope, Shepard, Shields, Spring Avenue, Stoddard, Webster, New Webster. (The "Branches" of the Clinton and other schools are included.)

Form C.

THE PHYSIQUE OF THE SCHOOL CHILDREN IN ST. LOUIS.

The parents or guardians of school children are requested to fill out the following blank, and to return it to the teacher on the next school day:

    4. Name of pupil...
    5. Place of birth..
    6. Age at nearest birthday......................................
    7. In what country was pupil's father born.....................
    8. In what country was pupil's mother born.....................
    9. Occupation of father...
  10. Number of pupil's sisters living..............; dead..................
  11. Number of pupil's brothers living............; dead..................
  12. Residence of pupil, No..........., ...Street.

On the second day, Form C having been returned, answers 5, 6, 7, 8, 9, 10, 11 and 12 were written. Some ignorant or prejudiced parents refused to answer the questions in Form C and in such cases the answers were obtained as far as possible from the school registers. While the first school was employed in this way on the second day, another school began to answer questions 1, 2, 3, 4 and 32. On the third day, the first school measured the height standing, height sitting and span of arms, while the second school was busy with the second day's work and the third school with the first.

Thus the thirty-two questions were divided into groups, and by the ninth day nine schools were working simultaneously, each on a different group. As soon as one school finished with an apparatus, it was taken to the next school on the list, almost always the nearest in point of distance. An extract from "The Chart of Days" will make this clear.

A PORTION OF "THE CHART OF DAYS."

NAME OF SCHOOL.	SPECIAL ASSISTANT.	RECORD:							
		Answers 1, 2, 3, 4, 13, 14, 32. (Color of Hair and Eyes.)	Answers 5, 6, 7, 8, 9, 10, 11, 12.	Height Standing, Height Sitting, Span of Arms.	Strength of Squeeze.	Girth of Chest.	Weight.	Acuteness of Vision.	Acuteness of Hearing.	Head Measurements.
Laclede	Lemen	4*	5	7	8	11	12	13	14	15
Madison and New Madison	Simmons	5	6	8	11	12	13	14	15	18
Pestalozzi	Newcomb	6	7	11	12	13	14	15	18	19
Carroll	Loth	7	8	12	13	14	15	18	19	20
Lafayette and Lafayette Branch	Taussig	8	11	13	14	15	18	19	20	21
Humboldt	Gooden	11	12	14	15	18	19	20	21	22
Lyon	Soper	12	13	15	18	19	20	21	22	25
Shepard	Blair	13	14	18	19	20	21	22	25	26
Meramec	Gross	14	15	19	20	21	22	25	26	27
Blow and Blow Branch	Schlossstein	15	18	20	21	22	25	26	27	....

The Laclede School began Jan. 4 and finished Jan. 15, the Madison School began Jan. 5 and finished Jan. 18, and so through the list.

Nearly nine thousand pupils were examined each school-day after the ninth, and each day saw the returns of nearly a thousand pupils completed. The collection of material was finished in fifty-four school-days, beginning Jan. 4 and ending March 18. During this period of about 11 weeks, 18059 girls and 16295 boys were examined† and nearly one million data collected, about five hundred thousand of which were measurements, a result which would hardly have been possible without the strenuous exertions of my assistants.

One of the Special Assistants mentioned in the Chart of Days was assigned to each school under investigation. It was his duty to visit the school every day for the purpose of giving any explanation which might be necessary. He was

---

* This and the following numbers are the days of the month on which the respective groups of measurements were made.

† Owing to absentees, the number of returns of any one measurement, e. g., height, was less than the number here given.

charged also to see that his school finished its allotted daily task and dispatched at the close of the session the apparatus in its hands to the next school on the list. In addition to these nine Special Assistants, thirty Head Measurers were employed. The Head Measurers were divided into five sections named after the days of the school week from Monday to Friday. Each section went on its own day to the school whose turn it was to have heads and faces measured. The remaining measurements were made by the teachers.

Not even the industry of these many assistants would have accomplished the task in the required time had there not been an abundance of apparatus. Of measuring-rods, measuring-tapes for the chest, double sets of Snellen-Green test letters, spectacle blinders, watches and callipers, there were a dozen each, and five dynamometers and six small platform-scales were also in constant use. Twelve school rooms could thus often work at the same time, and a great many children were measured in a few hours.

The following printed instructions were given to every measurer.

Form D.

INSTRUCTIONS TO OBSERVERS.

Four printed forms are furnished. These are:

  Form A. Male (white paper).
  " B. Female (tinted paper).
  " C. For parents.
  " D. Instructions to observers.

On Monday, January 4th, a sufficient number of these forms will be taken to school A. The same day answers 1, 2, 3, 4, 13, 14 and 32, Form A or B, are written, and each pupil is given Form C to carry home. The next morning form C is returned by the pupils, and answers 5, 6, 7, 8, 9, 10, 11 and 12 are copied on form A or B. The third day the measurements begin, and are carried out according to the following

CALENDAR.

  1st day 1, 2, 3, 4, 13, 14, 32.
  2d " 5, 6, 7, 8, 9, 10, 11, 12.
  3d " 15, 16, 17.
  4th " 18, 19.
  5th " 20, 21.
  6th " 22.
  7th " 23, 24.
  8th " 25, 26.
  9th " 27, 28, 29, 30, 31.

Form C and answers 1 to 12 in forms A and B, require no explanation. The remaining answers are to be sought by the following methods.

Arrange forms *A* and *B* in the order in which the pupils sit at their desks, and preserve this arrangement throughout the measurements.

13, 14. Color of Hair and Eyes.

Choose the adjective that most nearly indicates the color; cross out the rest with pen and ink.

15. Height Standing.

1. All the pupils unlace their shoes.
2. The teacher places the measuring rod against the cloak-room door casing, the projections on the rod serving to keep it parallel with the casing, and presumably perpendicular to the floor.
3. The pupil in the first seat comes forward, removes his shoes, stands on a folded newspaper upon which the rod also rests, his heels, body and head touching the door casing, the mouth closed and the chin somewhat depressed. Hair worn in a high knot must be let down.
4. Lower the sliding arm until the edge touches the crown of the pupil's head, and dictate the reading to a pupil assistant, who writes it opposite "15. Height standing." Meanwhile, the second pupil gets ready. [The rods are divided in centimetres and half centimetres; it will be easy to read to quarter centimetres. The dictation should be in the decimal system; thus: "One hundred fifteen, seventy-five (written 115.75), not one hundred fifteen and three quarters.]
5. The first pupil steps to one side, puts on his shoes, returns to his seat, and there laces his shoes. The second pupil removes his shoes and steps on the newspaper. The third pupil comes forward.

16. Height Sitting.

Place an armless wooden chair with a flat seat sideways against the door casing. The measuring rod is held perpendicularly on the seat, the projections on the rod touching if possible the casing. Pupils come forward as before.

Take care: 1, that the lower part of the spinal column touches the rod; 2, that the mouth is closed and the chin somewhat depressed; 3, that hair worn in a knot on the back of the head does not introduce an error. Measure, and dictate the reading.

17. Span of Arms.

1. Draw on the wall a chalk mark parallel with the floor and as high as the chin of a pupil of average height.
2. Hold the measuring rod parallel to the line and as high as the neck of the pupil to be measured.
3. The pupil touches one end with the middle finger of one hand and stretches along the rod as far as he can reach; chin up, heels together, body as close as possible to the rod.
4. Dictate the reading.

[In this and all other measurements the place of the pupil measured should be instantly taken by another.]

18, 19. Strength of Squeeze.

1. Depress the trigger of the dynamometer until the point of the indicator is exactly over the zero line of the scale.
2. The pupil grasps the oval ring in the right hand and squeezes his best.
3. Read the outer scale (graduated from 0 to 160) to pounds* as exactly as possible. Write the number after "18. Strength of Squeeze, right hand."
4. Reset the instrument, and test left hand. [Please do not touch the indicator. Always use the trigger.]

20, 21. Girth of Chest.

1. Take four pupils into the cloak room. They remove clothing over chest down to the garment next the skin. The measurements are made *on a level* with the nipples, and are dictated in centimetres and decimal fractions of a centimetre to a pupil assistant, who writes them opposite 20 and 21.
2. "Forced expiration." Pupil breathes out, makes chest as small as possible, inclines head forwards, draws shoulders slightly together.
3. "Forced inspiration." Shoulders back, head raised, deepest possible inspiration.

[Some children will require to be shown how to do this. As fast as a pupil is measured, dresses and returns to his seat, another pupil enters the cloak room to take his place, and strips for measurement.]

22. Weight.

The upper figures on the beam and the smaller of the two numbers on the iron weights are used with the scale pan. Pay no attention to them. The lower figures on the beam read from 1 to 45 pounds. Place the pupil's approximate weight on the counterpoise at the end of the beam; move the poise along the latter until the scale balances. Add the larger number on the hanging weight to the number marked on the lower scale by the poise. The sum is the pupil's weight.

23, 24. Acuteness of Vision.

It is necessary to make these tests between 10:30 a. m., or better 11 a. m., and 3 p. m., as at other hours the winter light is insufficient.

A pupil must not see the card before the moment for his test has come. Nor can the pupils who have been tested be allowed to communicate with those yet to be tested. Otherwise the letters will be committed to memory and the test spoiled.

1. Hang the test card on the wall opposite the windows, in a good light, and level with the pupil's eyes. Turn down the middle leaf of the card, so that only one set of test letters is visible.
2. Stand the pupil with toes touching a chalk mark 16 feet 5 inches from the card, and cover the left eye with the blinder furnished for that purpose. The head must be held straight, so that the child cannot see the test letters with the blinded eye.

---

* wo dynamometers were graduated metrically.

3. The child reads the test letters beginning with the largest; the smallest letter which can be read indicates the "acuteness of vision" at 16 feet 5 inches (5 metres). Mark the arabic figure under this letter opposite "Acuteness of Vision, right eye."

4. Change the blinder to the right eye.

5. Turn up the middle leaf of the test card so as to expose the second set of test letters.

6. Test the acuteness of vision of the left eye in a similar manner.

25, 26. *Acuteness of Hearing.*

The hearing tests are to be made in the cloak-room with the doors closed. Their success depends primarily on the absence of all noise from the adjacent class-rooms and halls.

1. The pupil is seated with the right ear towards the observer.

2. An assistant closes the pupil's left ear by pressing the *tragus* (the little cartilage in front of the external opening) inwards with her thumb while with the fingers of the same hand she closes the eyelids with a handkerchief. With the other hand the assistant holds the end of the brass ring of the tape measure against the head just beneath the ear, grasping the ring between thumb and forefinger and keeping the hand entirely below the auditory opening.

3. The observer stands directly opposite the ear and 11 feet from it, holding the tape parallel with the floor. The watch is held by the chain ring against the median line of the observer's body, just above and not touching the tape. The face of the watch is towards the pupil.

4. The observer charges the pupil to say "now" when he hears the watch tick, and advances the watch until the pupil replies. The watch is now withdrawn and again advanced until the pupil replies. Marking this point on the tape measure between finger and thumb, the observer puts the watch behind her back, and asks "Do you hear now?" If the pupil replies "No," the watch is returned to the former position and the inquiry repeated. If the watch is again heard at this distance the test is satisfactory; if not, it should be repeated. Write the distance at which the watch is heard in feet and inches opposite "Acuteness of Hearing, right ear."

5. Turn the pupil around (180°) and test the left ear in a similar manner.

The wilfulness of some children will make the accuracy of their tests doubtful. In such cases write an interrogation mark after the figures secured.

27, 28, 29, 30. *Head Measurements.*

The teacher carries the blanks arranged in the order in which the pupils sit in her room, a pen, and a centimetre rule. The medical assistant makes a measurement, places the callipers on the rule which the teacher holds, and reads the number of centimetres and millimetres. The teacher writes these numbers while the assistant makes the next measurement.

Instructions for Medical Assistants.

Measurements must be read to millimetres.

1. *Length of head.* — Place one point of the callipers on the most prominent point of the forehead, between the eyebrows. Bring the other point of the callipers down to the posterior part of the head and move it along the middle line until the greatest length of the head is found. (A, B.)

2. *Breadth of head.*— Take the greatest breadth of the head between the ears wherever it is found. Hold the callipers horizontally and perfectly symmetrically, approximately at F.

3. *Height of face.*— Put one point of the callipers in the deepest depression on the nose between the eyes (C)— then press the other point against the chin and find that point at D where the chin turns backwards. Ask the person to press his teeth together.

4. *Breadth of face.*— This is the greatest breadth between the narrow, bony ridges felt in front of the ears. The ridges run from the cheek bones to the ears. Hold the callipers horizontally and symmetrically, approximately at G.

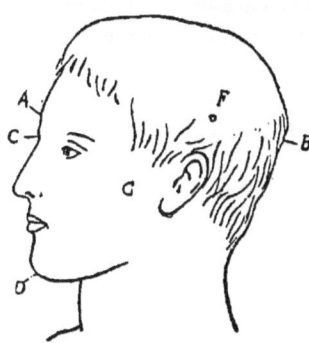

Position of the Points at which Head Measurements are Taken.
A-B, length of head. Approximately at F, breadth of head. C-D, Height of face. Approximately at G, breadth of face.*

5. *Hair line.*— This is the height of face from the point of the chin to the point where the hair begins to grow above the forehead.

Further information, if desired, may be had by addressing the Physiological Laboratory of Washington University, 625 Clark avenue.†

W. TOWNSEND PORTER.

Approved Jan. 4th, 1891.
   E. H. LONG,
      Superintendent.

Some changes were made in the above instructions during the course of the investigations. The tests of acuteness of hearing were found impracticable because of the unavoidable noise in the schools and were given up after about seven thousand pupils had been tested. It was noticed that the number possessing less than the normal acuteness of hearing was very large, and more than one pupil was found who had been punished for inattention, the result of an unsuspected

---

* I am indebted to Dr. Franz Boas for this figure.

† Since removed to 1806 Locust street. Dr. Porter's present address is Harvard Medical School, Boston, Mass.

deafness. Although the returns did not seem sufficiently trustworthy for statistical treatment, they justified the statement that not a little deafness exists unknown to both teacher and parents. This hidden infirmity deprives the pupil of much of the benefit of class-room instruction.

The measurement of the girth of chest over the garment next the skin was also modified after several thousand boys and girls had been measured, and the boys were made to remove coats and vests while the girls were measured over the indoor dress. The few girls who wore stays laid them aside.

CHAPTER II.

THE STATISTICAL METHODS EMPLOYED.

All measurements are accompanied by unavoidable errors. Thus the attempt to measure the height of a man is influenced by the accuracy with which the measuring apparatus is constructed, the care with which it is used, the position of the man's head on the vertebral column, the thickness of the intervertebral disks and a multitude of other factors. Some of these influences would make the observed height greater than the true height, others would make it less. The one group tends to counteract or compensate the other, and the result of their conflict is the measurement actually observed. The observed height therefore is never, except by chance, the real height, but deviates from it in one or the other direction — is now above and now below the truth — as one or the other group of influences gets the upper hand. The greater the number of influences, the more perfectly does compensation take place and the more nearly does the observed result approach the truth. But the truth itself can never be known, for only when the number of influences is infinite, can the probability of perfect compensation between them rise to a certainty. That which we call true is merely the probable truth and is worthy of confidence in exact proportion to its numerical probability.

The influences which affect a measurement are of two sorts, the one accidental and varying, such as, taking the measurement of height for an example, the degree of inclination of the head to the axis of the body, the placing of the measuring-rod and the like, the other constant and unvarying, such as an inaccuracy in the construction of the measuring-rod or a persistent bias in the mind of the observer. In both classes, the degree of compensation varies with the number of influences, for even a constant cause, although not accidental in its nature and found always on one side of the mean, may be compensated by another constant cause on the opposite side of the mean. The accidental or varying influences, however, are much more numerous than the constant or unvarying, and thus compensation is in their case much more complete. Constant, unvarying influences are therefore more deceptive than varying, accidental influences.

In practice, the influences determining the result of a single measurement elude all attempts at calculation, and the extent to which the result of a single observation approximates the truth cannot be told. But if many measurements are made of the same thing, for example the height of a man, the individual measurements arrange themselves on either side of the true height. If only accidental influences have been at work in each individual measurement in the series, the distribution of the observed heights on either side of the real height would, if the number of observations were infinite, be symmetrical. Thus the true height would at once appear, and the probability or degree of deviation of any single measurement would also be visible. Even with finite numbers a probability so great as to amount to a practical certainty can be secured. If 2,000 measurements of the same physical dimension, e. g. the height of an individual, are made carefully, the total number of observations will be distributed nearly as follows: —

TABLE No. 1.

THE DISTRIBUTION OF 2,000 MEASUREMENTS OF THE SAME QUANTITY WHEN THE DEVIATION OF THE INDIVIDUAL OBSERVATIONS FROM THE TRUE VALUE OF THE MEASURED QUANTITY IS DUE TO PURELY ACCIDENTAL CAUSES.*

$d =$ the Probable Deviation.	Number of Observations.
Greater than $+ 5.0\ d$	1
$+ 4.5\ d$ to $+ 5.0\ d$	1
$+ 4.0\ d$ " $+ 4.5\ d$	5
$+ 3.5\ d$ " $+ 4.0\ d$	11
$+ 3.0\ d$ " $+ 3.5\ d$	25
$+ 2.5\ d$ " $+ 3.0\ d$	49
$+ 2.0\ d$ " $+ 2.5\ d$	85
$+ 1.5\ d$ " $+ 2.0\ d$	135
$+ 1.0\ d$ " $+ 1.5\ d$	188
$+ 0.5\ d$ " $+ 1.0\ d$	236
$\pm\ .0$ " $+ 0.5\ d$	264
$- 0.5\ d$ " $\pm\ .0$	264
$- 1.0\ d$ " $- 0.5\ d$	236
$- 1.5\ d$ " $- 1.0\ d$	188
$- 2.0\ d$ " $- 1.5\ d$	135
$- 2.5\ d$ " $- 2.0\ d$	85
$- 3.0\ d$ " $- 2.5\ d$	49
$- 3.5\ d$ " $- 3.0\ d$	25
$- 4.0\ d$ " $- 3.5\ d$	11
$- 4.5\ d$ " $- 4.0\ d$	5
$- 5.0\ d$ " $- 4.5\ d$	1
Greater than $- 5.0\ d$	1

This result is quite independent of the nature of the thing measured, provided that no unvarying influences are at work.

An examination of Table No. 1 shows (1) that the distribution of the observations is symmetrical about a median point, in other words that equal deviations to one and the other side of this point are equally probable; (2) that half the whole number of observations fall within $\pm\ 1.0\ d$ of the most probable value, indicating that a small deviation is more probable than a large one; and (3) that there is a limit beyond which no deviation occurs. Such are the peculiarities of a series of observations in which the deviations from the true value are due to purely accidental causes. It is evident that the most probable value of the true measurement is typical of the whole series, and that the degree of probability of any devia-

---

* After Thoma: Untersuchungen über die Grösse der anatomischen Bestandtheile des menschlichen Körpers im gesunden und kranken Zustande. Leipzig, 1882. Page 28.

tion from the type is easily calculated; for example in the series of Table No. 1, the chances are even that any deviation will fall between $+1.0\,d$ and $-1.0\,d$, for it has been seen that half the whole number of observations fall within these limits.

Quetelet demonstrated that the method of grouping related measurements could be used not only for showing the most probable height of one individual but also for showing the most probable or typical height, i. e., that most often found, of a number of individuals. The Belgian Astronomer Royal pointed out that the series obtained by measuring a number of individuals of the same type, e. g., men of the same nationality, was characterized by the peculiarities that distinguish a series in which the deviations from the typical or most probable value are due to accidental causes. The truth of this statement is obvious when the two are compared.

TABLE No. 2.*
HEIGHTS OF UNITED STATES RECRUITS.

Height at Intervals of One Inch.	Number of Recruits.
Between 78 and 79 inches	2
" 77 " 78 "	6
" 76 " 77 "	9
" 75 " 76 "	42
" 74 " 75 "	118
" 73 " 74 "	343
" 72 " 73 "	680
" 71 " 72 "	1485
" 70 " 71 "	2075
" 69 " 70 "	3133
" 68 " 69 "	3631
" 67 " 68 "	4054
" 66 " 67 "	3475
" 65 " 66 "	3019
" 64 " 65 "	1947
" 63 " 64 "	1237
" 62 " 63 "	526
" 61 " 62 "	50
" 60 " 61 "	15
" 59 " 60 "	10
" 58 " 59 "	6
" 57 " 58 "	7
" 56 " 57 "	3
" 55 " 56 "	1
" 55 and less	4
Total...	25878

* E. B. Elliott's table in Medical Statistics of the Provost-Marshal-General's Bureau, by J. H. Baxter, 1875, Vol. I. Introduction, page lxxx.

If the individual heights of a group of adults are found to be approximations of a middle, typical height, it would seem that the heights of children of the same sex, age and class must show a similar relation to a type; and this inference is justified by observation.

TABLE No. 3.
OBSERVED DISTRIBUTION OF THE HEIGHTS OF 2192 ST. LOUIS SCHOOL GIRLS, AGED 8.

Heights at Intervals of 2 Centimetres.	Number of Observations.
141 and 142 cm.	1
139 " 140 "	
137 " 138 "	1
135 " 136 "	5
133 " 134 "	10
131 " 132 "	21
129 " 130 "	28
127 " 128 "	79
125 " 126 "	138
123 " 124 "	183
121 " 122 "	243
119 " 120 "	342
117 " 118 "	321
115 " 116 "	297
113 " 114 "	222
111 " 112 "	137
109 " 110 "	84
107 " 108 "	42
105 " 106 "	27
103 " 104 "	8
101 " 102 "	2
99 " 100 "	1
Total	2192

The characteristics of a series in which the individual observations are accidental deviations from a typical middle value are established by methods which have long been used by astronomers and mathematicians. These methods seek to determine (1) the middle or typical value and (2) the way in which the individual observations are dispersed on either side of this value. Some statisticians take the average to be the nearest approach to the typical value, others prefer the mean, i. e., median value.* It will be necessary to consider at some length the relative value of average and mean in anthropometrical studies of the growth of children, but for the present a

---

* Throughout this work the word MEAN is used as the synonym of MEDIAN VALUE, and not in the sense of ARITHMETICAL MEAN, which is called here the AVERAGE. The MIDDLE VALUE will be employed as meaning either Median or Average value.

statement of the methods by which the mean and average are secured will suffice.

By AVERAGE (A) is meant the quotient obtained by dividing the sum (Σa) of the values (a) obtained in the individual measurements by the whole number of measurements (n).

$$A = \frac{\Sigma a}{n} \qquad (1)$$

The calculation of the average height of girls, aged 9, will serve as an example.

TABLE No. 4.

THE CALCULATION OF THE AVERAGE HEIGHT OF ST. LOUIS SCHOOL GIRLS, AGED 9.

Height in Centimetres.	Number of Observations. (n)	Product. (a)	Height in Centimetres.	Number of Observations. (n)	Product. (a);
168	1	168	134	25	3350
167			133	31	4123
166			132	29	3828
165			131	62	8122
164			130	78	10140
163			129	91	11739
162			128	119	15232
161			127	132	16764
160	1	160	126	143	18018
159			125	162	20250
158			124	141	17484
157			123	162	19926
156			122	145	17690
155			121	130	15730
154			120	145	17400
153			119	97	11543
152			118	94	11092
151			117	69	8073
150			116	59	6844
149			115	47	5405
148			114	30	3420
147			113	18	2034
146			112	11	1232
145			111	10	1110
144			110	10	1100
143	1	143	109	5	545
142	2	284	108	1	108
141	3	423	107	5	535
140	2	280	106	3	318
139	3	417	105	1	105
138	9	1242	104	1	104
137	12	1644	103	1	103
136	15	2040	102		
135	14	1890	101		
			100	2	200
			Total.....	$n = 2122$	$\Sigma a = 262358$

$$A = \frac{\Sigma a}{n} = \frac{262358}{2122} = 123.64 \text{ cm.}$$

If the number of observations is very large, the MEAN (MEDIAN value, M,) may be found with considerable accuracy by a mere inspection of the series. The mean of such a series is the measurement which most frequently recurs. Thus, the mean height of the recruits of Table No. 2, page 278, is between 67 and 68 inches. The accuracy with which the mean can thus be found depends not only on the number of observations, but also on the size of the units of measurement.

For most purposes it is desirable to know not merely at which inch, centimetre, kilogramme or other unit the greatest number of observations is found, but exactly at what fraction of the unit. Again, the relation between the number of observations and the size of the unit may be such that the largest number of observations at one unit will not fall at the true mean, or line dividing the total number of observations into two equal groups. The method by which the mean can be calculated with exactness will be illustrated by the following example. The mean height of the girls in Table No. 4 is obtained by adding the number of observations from below upwards until the sum cannot be increased by the next number in the column without exceeding half of the total number of observations. Thus 1046 is reached opposite 123 cm.; the next number in the column (141) would make the sum 1187, which is more than the half (1061) of the total number of observations (2122). The mean is, therefore, greater than 123 cm. but less than 125 cm. Its position is found by interpolation. Half of the total number of observations is 1061, which is 15 more than the sum of the observations up to 124 cm.; 15 is 11 per cent. of 141, the observations at 124 cm. Hence, the mean is 124.15 cm.

Neither the mean nor the average can give any information as to the way in which the individual observations in a series are distributed, and it is plain that two series having an identical mean or average may differ greatly in respect of the dispersion of the individuals from the middle value. Thus the two very different series —

$$4, 5, 6, 14, 15, 16$$
$$9, 9, 10, 10, 11, 11$$

have the same average (10). The best measure of the degree of dispersion or deviation of the individual members of a series from their common mean or average is that afforded by the Probable Deviation. The probable deviation is that deviation from the middle value, which, in a large series of observations, is as often exceeded as attained (Lexis). In other words, half of the whole number of observations fall short of the probable deviation, while the other half exceed it. A deviation which exceeds is as probable as one that does not reach this value. The probable deviation can be calculated with the formula —

$$d = \pm\, 0.6745 \sqrt{\frac{\Sigma \delta^2}{n-1}} \qquad (2)$$

In this formula:

d = the probable deviation.
$\Sigma \delta^2$ = the sum of the squares of the individual deviations from the mean or average.
n = the number of observations in the series.

In the place of this formula, in which the calculation of $\Sigma \delta^2$ requires much labor when the number of observations is great, a simpler formula may be safely used in getting the deviation in a large series.

$$d = \pm\, 0.8453 \frac{\Sigma \delta}{n} \qquad (3)$$

The formula says that all the individual deviations from the mean or average of a series must be added together without regard to whether they are plus or minus and divided by the total number of observations. The working of the formula will be illustrated by finding the probable deviation from the average height of girls aged 8.

TABLE No. 5.

The Calculation of the Probable Deviation (d) from the Average Height (118.36 Cm.) of 2193 Girls, aged 8.

Height at Intervals of 2 Centimetres.	ν	δ	$\nu\delta$
141 and 142 Cm.	1	23.64	23.64
139 " 140 "			
137 " 138 "	1	19.64	19.64
135 " 136 "	5	17.64	88.20
133 " 134 "	10	15.64	156.40
131 " 132 "	21	13.64	286.44
129 " 130 "	28	11.64	325.92
127 " 128 "	79	9.64	761.56
125 " 126 "	138	7.64	1054.32
123 " 124 "	183	5.64	1032.12
121 " 122 "	243	3.64	884.52
119 " 120 "	342	1.64	560.88
117 " 118 "	321	0.36	115.56
115 " 116 "	297	2.36	700.92
113 " 114 "	222	4.36	967.92
111 " 112 "	137	6.36	872.69
109 " 110 "	84	8.36	702.24
107 " 108 "	42	10.36	435.12
105 " 106 "	27	12.36	333.72
103 " 104 "	8	14.36	114.88
101 " 102 "	2	16.36	32.72
99 " 100 "	1	18.36	18.36
Total......	2192		9487.77

$$d = \pm\, 0.8453 \frac{9487.77}{2192} = \pm\, 3.698 \text{ Cm.}$$

The distribution of the above series of the heights of girls, aged 8, is therefore characterized by a probable deviation of ± 3.7 cm.; that is, one against one may be wagered that no girl aged 8 will be taller than 122.06 cm. or shorter than 114.66 cm. If the number of observations falling between $A \pm d$, $A \pm 2d$, $A \pm 3d$...... $A \pm {}^nd$ be noted, a complete picture of the individual observations in a series will be obtained. This observed distribution may then be compared directly with the distribution of the observations in an hypothetical series constructed according to the calculus of probabilities. The observed and the theoretical series should correspond, if the causes of deviation are purely accidental. It has already been said that such a comparison must be made before it can be known whether the observations in any series can be treated by the methods of the theory of probabilities. It is however not necessary to compare more than one of a number of series in which the distribution follows a common type. It will suffice for our purposes therefore to compare the distribution of the 2192 heights in Table No. 3, page 279, with the distribution of an equal number according to the calculus of probabilities. The method by which this is done permits the construction of a curve from the observations actually made which shall be the probable form of a curve representing the distribution of the entire class from which the observations have been drawn.

The number of observation (Z) which should be included between the average (A) and any deviation from the average, in other words any multiple (m) of the probable deviation (d), is obtained from the equation.*

$$P = \frac{2}{\sqrt{\pi}} \int_0^t e^{-t^2} dt \qquad (4)$$

The labor of calculating deviations with the aid of this equation is avoided by the use of such tables as Stieda's, reproduced below.

TABLE No. 6.

STIEDA'S TABLE FOR CALCULATING THE NUMBER OF OBSERVATIONS AT ANY DISTANCE FROM THE MEAN OR AVERAGE WITHIN THE LIMITS: M+ 5d AND M —5d.

p	Percent.	p	Percent.
0.1	5.4	1.8	77.5
0.2	10.7	1.9	80.0
0.3	16.0	2.0	82.3
0.4	21.3	2.1	84.3
0.5	26.4	2.2	86.2
0.6	31.4	2.3	87.9
0.7	36.3	2.4	89.5
0.8	41.1	2.5	90.8
0.9	45.6	2.6	92.1
1.0	50.0	2.7	93.1
1.1	54.2	2.8	94.1
1.2	58.2	2.9	95.0
1.3	61.9	3.0	95.7
1.4	65.5	3.5	98.2
1.5	68.8	4.0	99.3
1.6	71.9	4.5	99.8
1.7	74.8	5.0	99.93

* Kramp. L'Analyse des réfractions astronomiques.

If p is allowed to represent one of the figures in the first column and A the average of all the measurements in a series, the figures opposite p in the second column will give the per cent. of individual measurements lying within the limits:

$$A + p \cdot d \text{ and } A - p \cdot d$$

Suppose for example it were required to know how many of a series of 2192 girls aged 8 were of a height between the average (118.36 cm.) and a deviation of $\pm 1.5\, d$ (1.5. 3.7 cm. = 5.55 cm.), i. e. between 118.36 cm. + 5.55 cm. = 123.91 cm. and 118.36 cm. — 5.55 cm. = 112.81 cm. The number in the table opposite 1.5 is 68.8, which says that 68.8 per cent. of the 2192, or 1508, should fall within the limits stated. Then half this number must fall between A and A + 1.5 d (118.36 cm. and 123.91 cm.). In a similar manner it will be found that 50 per cent. of the whole number, or 1096, should fall within the limits A $\pm$ d (118.36 cm. + 3.7 cm. = 122.06 cm. and 118.36 cm. — 3.7 cm. = 116.51 cm.), and 25 per cent. between A and A + d (118.36 cm. and 122.06 cm.). Thus may be calculated the number of observations which should occur at any deviation from the average. The theoretical and observed distribution of 2192 girls, aged 8, is compared in Table No. 7.

TABLE No. 7.

The Theoretical and the Observed Distribution of the Heights of 2192 Girls, Aged 8.

Probable Deviation d.	Heights at Intervals of ± 0.5d.	Theoretical Distribution.	Observed Distribution.
+ 5.0 d	136.86 Cm.	2	2
+ 4.5 "	135.01 "	6	8
+ 4.0 "	133.26 "	12	18
+ 3.5 "	131.31 "	27	27
+ 3.0 "	129.46 "	54	57
+ 2.5 "	127.61 "	93	104
+ 2.0 "	125.76 "	148	150
+ 1.5 "	123.91 "	206	209
+ 1.0 "	122.06 "	259	286
+ 0.5 "	120.21 "	289	300
0.0 "	118.36 "		
− 0.5 "	116.51 "	289	272
− 1.0 "	114.66 "	259	275
− 1.5 "	112.81 "	206	196
− 2.0 "	110.96 "	148	126
− 2.5 "	109.11 "	93	80
− 3.0 "	107.26 "	54	40
− 3.5 "	105.41 "	27	26
− 4.0 "	103.56 "	12	12
− 4.5 "	101.71 "	6	3
− 5.0 "	99.86 "	2	2
Total......		2192	2192

The curves in Plate I* give a graphic representation of the figures in Table No. 7. A glance at these curves shows that there is a close agreement between them, indicating that the individual observations out of which they are constructed are symmetrically grouped about a middle value typical of the whole, and proving that the material with which we have to deal satisfies the requirements of theory.

The method of showing distribution just described is not the only method used for this purpose. Francis Galton † determines the actual distribution of the observations in a series at intervals of 5 or 10 per cent. from the median or 50 "percentile grade." The calculation of the median value (mean) or 50 percentile grade has already been described: that for the 5, 10, 20 and other percentile grades is similar in

---

* The plates are placed after the index at the end of the number.
† Natural Inheritance. London, 1889.

principle. The percentile grades of the heights of girls, aged 9, are as follows: —

TABLE No. 8.

THE PERCENTILE DISTRIBUTION OF THE HEIGHTS OF GIRLS, AGED 9.

Percentile Grades.	Number of Observations.	Heights.
95	2015.9	133.42 Cm.
90	1909.8	130.97 "
80	1697.6	128.62 "
75	1591.5	127.58 "
70	1485.4	126.54 "
60	1273.2	125.53 "
50	1061.0	124.11 "
40	848.8	122.76 "
30	636.6	121.21 "
25	530.5	120.90 "
20	424.4	119.59 "
10	212.2	117.12 "
5	106.1	115.17 "
Total............	2122	

The Probable Error of the Average can be determined by means of the formula

$$E = \pm \frac{d}{\sqrt{n}} \qquad (5)^*$$

where E = the probable error of the average,
  d = the probable deviation of an individual from the average,
  n = the number of observations in the series.

Substituting the values determined for the heights of girls, aged 9, we have —

$$E = \pm \frac{3.698}{\sqrt{2122}} = \pm 0.079 \text{ cm}.$$

The values of E are given in Table No. 9.

---

* Formulas (1), (2), (3) and (5) and Table No. 6 are taken from L. Stieda's admirable article: Ueber die Anwendung der Wahrscheinlichkeitsrechnung in der anthropologischen Statistik. Archiv. für Anthropologie, Bd. xiv., 1882, p. 167-182.

TABLE No. 9.

THE PROBABLE ERROR OF THE AVERAGE: $E = \pm \dfrac{d}{\sqrt{n}}$, where e = Probable Error, d = Probable Deviation, n = Number of Observations.

Dimensions.	Sex.	Unit of Measurement.	Age at Nearest Birthday, and Probable Error.															
			6	7	8	9	10	11	12	13	14	15	16	17	18	19	20	21
Weight, in Indoor Dress.	Boys.	Kilogramme	±0.054	±0.039	±0.042	±0.015	±0.049	±0.062	±0.051	±0.110	±0.148	±0.227	±0.431	±0.521	±	±	±	±
	Girls.		0.051	0.045	0.042	0.049	0.052	0.070	0.081	0.115	0.142	0.154	0.207	0.244	0.289	0.332	0.565	
Height Standing, Without Shoes.	Boys.	Centimetre.	±0.128	±0.084	±0.082	±0.080	±0.087	±0.099	±0.116	±0.140	±0.153	±0.286	±0.427	±0.592	±0.924			0.651
	Girls.		0.123	0.089	0.079	0.083	0.089	0.106	0.098	0.150	0.156	0.151	0.197	0.241	0.265	0.438	0.353	
Height Sitting	Boys.	Centimetre.	±0.105	±0.061	±0.048	±0.049	±0.053	±0.060	±0.067	±0.076	±0.103	±0.161	±0.250	±0.427	±0.536			0.290
	Girls.		0.132	0.053	0.044	0.046	0.049	0.057	0.063	0.078	0.095	0.098	0.116	0.153	0.133	0.198	0.230	
Span of Arms	Boys.	Centimetre.	±0.144	±0.096	±0.088	±0.089	±0.103	±0.113	±0.112	±0.159	±0.197	±0.321	±0.574	±0.581	±0.761			0.712
	Girls.		0.140	0.107	0.092	0.092	0.104	0.116	0.109	0.150	0.160	0.176	0.217	0.285	0.334	0.517	0.474	
Girth of Chest, Inspiration.	Boys.	Centimetre.	±0.083	±0.057	±0.051	±0.053	±0.059	±0.045	±0.072	±0.089	±0.116	±0.172	±0.305	±0.382	±0.551			0.482
	Girls.		0.087	0.061	0.053	0.056	0.060	0.072	0.078	0.097	0.113	0.152	0.152	0.217	0.243	0.340	0.331	
Girth of Chest, Expiration.	Boys.	Centimetre.	±0.083	±0.057	±0.052	±0.056	±0.063	±0.061	±0.076	±0.088	±0.119	±0.165	±0.281	±0.374	±0.504			0.560
	Girls.		0.094	0.062	0.053	0.058	0.062	0.078	0.083	0.098	0.115	0.148	0.175	0.248	0.265	0.420	0.347	
Strength of Squeeze, Right Hand.	Boys.	Kilogramme	±0.056	±0.039	±0.045	±0.055	±0.061	±0.068	±0.082	±0.126	±0.143	±0.244	±0.414	±0.458				
	Girls.		0.053	0.043	0.043	0.049	0.053	0.060	0.072	0.093	0.133	0.147	0.200					
Strength of Squeeze, Left Hand.	Boys.	Kilogramme	±0.057	±0.044	±0.048	±0.059	±0.059	±0.067	±0.080	±0.103	±0.134	±0.243	±0.438	±0.639				
	Girls.		0.056	0.042	0.046	0.049	0.053	0.059	0.076	0.094	0.127	0.154	0.207	0.264				
Length of Head	Boys.	Millimetre.	±0.171	±0.118	±0.106	±0.109	±0.094	±0.129	±0.115	±0.131	±0.182	±0.208	±0.355	±0.575	±0.717		0.433	
	Girls.		0.170	0.108	0.092	0.101	0.102	0.119	0.114	0.132	0.143	0.180	0.256	0.275	0.252	0.466		
Width of Head	Boys.	Millimetre.	±0.118	±0.090	±0.082	±0.084	±0.090	±0.095	±0.092	±0.114	±0.129	±0.186	±0.249	±0.486	±0.685		0.426	
	Girls.		0.154	0.099	0.082	0.083	0.093	0.097	0.105	0.115	0.130	0.150	0.210	0.242	0.300	0.466		
Root of Nose to Point of Chin.	Boys.	Millimetre.	±0.207	±0.120	±0.103	±0.107	±0.110	±0.127	±0.132	±0.160	±0.174	±0.238	±0.346	±0.744	±0.883			0.627
	Girls.		0.194	0.115	0.099	0.103	0.109	0.116	0.129	0.133	0.156	0.192	0.245	0.346	0.297	0.421	0.469	
Width of Face	Boys.	Millimetre.	±0.172	±0.101	±0.090	±0.101	±0.092	±0.108	±0.113	±0.122	±0.153	±0.196	±0.360	±0.627	±0.718			0.528
	Girls.		0.174	0.101	0.087	0.092	0.127	0.107	0.109	0.127	0.135	0.157	0.201	0.275	0.265	0.420	0.419	
Hair Line to Point of Chin.	Boys.	Millimetre.	±0.209	±0.159	±0.122	±0.127	±0.127	±0.143	±0.137	±0.173	±0.236	±0.292	±0.425	±0.853	±0.139			0.786
	Girls.		0.238	0.137	0.119	0.121	0.131	0.143	0.148	0.173	0.193	0.211	0.286	0.437	0.520	0.749	0.560	

The mean or average of the observations at any age in the period of growth is typical of the child at that age, and a comparison of the means at different ages will reveal the law of growth of the type. Again, the mean of the observations at any deviation from the mean of the whole number, for example the height at a deviation of $+ d$ from the mean, or, if Galton's method is employed, the height at any percentile grade, is the type of those who stand at a certain degree of deviation from the type of the whole number. Thus types of tall and short, light and heavy children are secured. The types of the same degree of deviation from the mean at all ages are as comparable as the type of the whole number of observations, and reveal the growth of the typically tall and short, light and heavy children; but the comparison is less secure the greater the deviation from the mean, for the probable error is inversely as the square of the number of observations, and the number of observations rapidly diminishes on either side of the mean.

The methods described in this chapter have been employed in the present investigation. For every entire series here presented, the mean and the average, the probable deviation and the probable error, and the 5, 10, 20, 25, 30, 40, 50, 60, 70, 75, 80, 90 and 95 percentile grades have been calculated. The 25 and 75 percentile grades were obtained by dividing by 2 the sums of the 20 and 30, and the 70 and 80 percentile grades respectively.

CHAPTER III.

TRUSTWORTHINESS OF THE MATERIAL; ITS LIMITATIONS. COMPARISON OF MEDIAN AND AVERAGE VALUES.

The question first to be decided in the discussion of an anthropometrical series is whether the individual observations are so related one to another that they constitute accidental deviations from a middle value. The method of answering this question by comparing the series of observations with a series constructed according to the theory of probabilities has already been described, and it has been stated incidentally that the heights of St. Louis girls aged 8 agreed with the theoretical series in their distribution. Where such an agreement exists, the individual observations are to be regarded as approximations of a middle value which is the type of the series. It is not necessary to make this comparison at more than one age, or in more than one dimension, for it is known that if one series in a group like that with which we have to deal shows this agreement the other series will be found to do the same. In the present case, additional evidence of the correspondence between observation and theory could be furnished, were it required, by Table No. 10, containing the probable deviation

TABLE No. 10.

THE PROBABLE DEVIATION (d) FROM THE AVERAGE: $d = \pm\, 0.8453 \cdot \dfrac{\Sigma \delta}{n}$, where

δ = Deviation from Average.
$\Sigma \delta$ = Sum of Individual Deviations.
n = Total Number of Observations.

Dimensions.	Sex.	Unit of Measurement.	Age at Nearest Birthday and Probable Deviation.															
			6	7	8	9	10	11	12	13	14	15	16	17	18	19	20	21
Weight.............	Boys. Girls.	Kilogramme	±1.43 1.44	±1.68 1.88	±1.96 1.95	±2.09 2.23	±2.23 2.31	±2.60 2.91	±2.46 3.31	±3.88 4.22	±4.56 4.67	±5.06 4.05	±6.16 4.24	±4.38 3.70	±3.60	±3.76	±3.76	
Height Standing....	Boys. Girls.	Centimetre.	3.40 3.42	3.61 3.75	3.89 3.70	3.75 3.83	3.98 4.06	4.23 4.48	4.47 5.23	4.98 5.46	5.58 5.15	6.33 4.01	5.87 4.05	5.15 3.45	4.98 3.39	4.04	3.08	4.27
Height Sitting.......	Boys. Girls.	Centimetre.	2.82 2.03	2.64 2.19	2.26 2.04	2.34 2.11	2.42 2.19	2.56 2.37	2.72 2.61	2.74 2.87	3.15 3.11	3.59 2.54	3.48 2.36	3.77 2.17	2.89 1.72	1.82	2.03	1.86
Span of Arms.......	Boys. Girls.	Centimetre.	3.35 3.87	4.16 4.18	4.18 4.28	4.25 4.18	4.70 4.59	4.84 4.87	1.57 4.51	5.71 5.65	6.03 5.29	7.15 4.68	7.89 4.41	5.03 4.05	4.31 4.28	4.71	4.13	4.33
Girth of Chest, Full Inspiration.	Boys. Girls.	Centimetre.	2.28 2.38	2.39 2.43	2.34 2.40	2.42 2.50	2.63 2.62	2.70 2.91	2.86 3.14	3.13 3.51	3.51 3.62	3.84 3.53	4.37 3.03	2.95 3.15	3.07 3.11	3.08	2.67	2.73
Girth of Chest Full Expiration.	Boys. Girls.	Centimetre.	2.15 2.57	2.37 2.51	2.36 2.40	2.60 2.56	2.80 2.71	2.52 3.18	3.01 3.33	3.08 3.56	3.64 3.67	3.69 3.80	4.01 3.51	2.35 3.52	2.81 3.35	3.78	2.84	2.91
Strength of Squeeze, Right Hand.....	Boys. Girls.	Kilogramme	1.41 1.39	1.52 1.67	1.95 1.87	2.45 2.11	2.66 2.27	2.74 2.42	3.17 2.83	4.27 3.31	4.15 4.11	5.10 3.65	5.29 3.77	4.87 3.17				
Strength of Squeeze, Left Hand.....	Boys. Girls.	Kilogramme	1.43 1.47	1.72 1.62	2.08 2.00	2.66 2.10	2.57 2.23	2.70 2.36	3.11 2.98	3.50 3.35	3.93 3.90	5.10 3.84	5.58 3.93	5.08 3.17				
Length of Head.....	Boys. Girls.	Millimetre..	4.21 4.20	4.57 4.22	4.85 4.26	4.87 4.39	4.13 4.33	5.26 4.70	4.56 4.46	4.55 4.55	5.44 4.53	4.66 4.58	4.93 5.12	5.08 4.09	4.13 3.20	4.09	3.75	3.66
Width of Head.....	Boys. Girls.	Millimetre..	2.82 3.81	3.58 3.86	3.68 3.66	3.74 3.65	3.85 3.96	3.76 3.80	3.61 4.03	3.90 3.92	3.82 3.96	4.03 3.84	3.48 4.17	4.21 3.60	3.88 3.85	4.14	3.64	3.08
Root of Nose to Point of Chin	Boys. Girls.	Millimetre..	5.09 4.80	4.80 4.45	4.67 4.46	4.82 4.48	4.77 4.64	5.18 4.61	5.22 5.05	5.54 4.68	5.21 4.94	5.21 4.98	4.79 4.91	6.57 5.15	4.92 3.79	3.77	4.01	3.66
Width of Face.....	Boys. Girls.	Millimetre..	4.24 4.30	4.09 3.92	4.08 3.97	4.44 4.02	4.05 4.38	4.52 4.17	4.42 4.26	4.27 4.45	4.56 4.26	4.33 4.08	5.01 4.07	4.69 4.08	4.13 3.38	3.76	3.58	3.08
Hair Line to Point of Chin.	Boys. Girls.	Millimetre..	5.18 5.88	6.40 5.28	5.46 5.27	5.67 5.38	5.55 5.62	5.80 5.70	5.44 5.76	6.01 6.04	7.05 6.11	6.41 5.40	5.90 5.69	7.60 6.20	8.40 6.13	6.22	4.76	4.52

from the average. It is seen in this table that the probable deviations are small, that is, one-half of the observations deviate but little from the middle value, which is one of the fundamental attributes of deviations due to accidental causes. But additional evidence is hardly required, and few critics will object to regarding the middle values in this investigation as types of their respective series.

The objection sometimes made that the errors of observation materially affect the truth of the values obtained is of little weight, partly because such errors are "accidental" and compensate each other as already explained, and partly because a deviation from the middle value due to an uncompensated error in measurement forms, as a rule, an inconsiderable part of that greater deviation which expresses the physiological difference between the individual and the type of his age and class. Accidental errors of observation need not give concern in measurements of great numbers of school children. Nor need there be much fear of constant errors of observation, provided the collection of material is made by many persons and with a good number of each sort of measuring instrument. If hundreds of teachers take part in the measurements, as in the present investigation, a constant cause of error due to a teacher's unconscious bias or personal equation on one side of the middle value will very probably be compensated by the bias of another teacher on the opposite side, and, similarly, if a number of scales are used, the errors of those which weigh too lightly are likely to be compensated by the errors of those whose readings are too heavy.

The trustworthiness of this material must be encouraging to those whose hold on the theory underlying these matters is not very strong, because it illustrates the truth that the types of physical development, and the laws of growth of the type may be induced from measurements made by comparatively unskilled hands and demonstrates that a system of anthropometrical measurements may be fruitfully employed in the public schools.

The use to which middle values and the deviations from middle values shall be put is in part the subject of controversy, and it will be well to state here the manner in which they shall be employed in the present work. In doing so it will be necessary to give a brief account of the matters concerning which agreement is general as well as those which are in dispute.

It is acknowledged generally that the method of Quetelet furnishes a middle value typical of the series from which it was drawn, for example the middle weight of boys of the same age, nationality and social condition is the typical weight of boys of that age and class; and it is further acknowledged that the increase in the middle value from year to year expresses the law of growth of the type. It follows that the middle value of those who stand at any deviation from the middle value of the whole number is the type of that degree of deviation from the type of the whole number and that the increase in the middle value at the same degree of deviation at each age in the period of growth expresses the law of growth of the type at that degree of deviation. Thus the curves of percentile grades printed below express the growth of the typical St. Louis school-boy and girl. The type at a certain deviation from the mean of an age will show the same degree of deviation from the mean at any subsequent age; for example a type-boy in the 75 percentile grade at age 6 will throughout his growth be heavier than 75 per cent. of boys of his own age. Percentile curves are of course not necessarily parallel. The type of the 50 per cent. who exceed the middle value of the whole number has a law of growth characteristic of tall boys and different from that of the type of the 50 per cent. who fall below the middle value of the whole number.

The application of the middle value to individuals has not yet been mentioned. It is here that controversy finds its hold. The relation of the individual to the type is not known. It is not known whether a boy who at age 6 is heavier than 75 per cent. of boys at his age will at age 18 be heavier than 75 per cent. of boys at that age. Some anthropologists believe that there is at least probability that children remain in the same percentile grade throughout life, while others dissent from this view. In truth the development of the individual has been little studied, and it is therefore not yet possible to state his probable future growth.

This question — the relation of the growth of the individual to the growth of the type — can be determined only by the individualizing method. The generalizing method deals solely with units, irrespective of their individuality. If for example John Smith in the 75 percentile grade and William Harrison in the 60 percentile grade at age 6 exchange places at age 7, the personal curve of each boy undergoes an important deviation, but the number of units in the two grades, and consequently the middle value in each, is unaltered. The generalizing method, therefore, furnishes no data by which the future development of individuals can be safely judged. The individualizing method, on the contrary, follows the individual from year to year throughout his growth and establishes the frequency and extent of his deviations from the growth of the type. The lack of data collected by the individualizing method is regrettable, but this gap in our knowledge does not prevent the establishing of physical standards by which the probability that the physique of any child is normal or abnormal can be fixed.

The facts stated above limit the application of middle values to (1) the establishing of physical standards at each age, and (2) the using of these standards to determine whether the physique of any child is normal: they do not, in the present state of knowledge, permit prediction of future growth.

It has been much disputed whether the median value or the average should be taken for the type. Many investigators agree with Sir John Herschel, who declares in his celebrated review of Quetelet's *Lettres sur la théorie des probabilités* (page 23) that an average " may be convenient, to convey a " general notion of the things averaged, but involves no con- " ception of a natural and recognizable central magnitude all " differences from which ought to be regarded as deviations " from a standard. The notion of a mean, on the other " hand, does imply such a conception, standing distinguished " from an average by this very feature, viz.: the regular " march of the groups, increasing to a maximum, and thence " again diminishing. An average gives us no assurance that " the future will be like the past. A mean may be reckoned " on with the most implicit confidence. All the phil-

"osophical value of statistical results depends on a due appreciation of this distinction and acceptance of its consequences."

Other statisticians have used the average exclusively. Out of respect to this difference of opinion, both the median value and the average have been employed in the present work. On calculating these values for each series, it was found that the difference between them was inconsiderable, showing that

TABLE No. 11.
MEDIAN MINUS AVERAGE VALUES.

Dimensions.	Sex.	Unit of Measurement.	Age at Nearest Birthday, and Median Minus Average Values.						
			6	7	8	9	10	11	12
Weight............	Boys.	Kilogramme	+0.10	−0.01	+0.09	+0.16	+0.04	+0.08	−0.10
	Girls.		+0.06	−0.01	−0.03	−0.04	−0.04	−0.35	−0.41
Height Standing....	Boys.	Centimetre..	+0.29	+0.45	+0.64	+0.52	+0.68	+0.60	+0.36
	Girls.		+0.43	+0.49	+0.38	+0.44	+0.42	+0.41	+0.43
Height Sitting......	Boys.	Centimetre..	−0.54	−0.10	+0.47	+0.35	−0.40	+0.42	+0.13
	Girls.		+0.66	+0.64	+0.99	+0.03	+0.69	+0.59	+0.55
Span of Arms......	Boys.	Centimetre..	+0.62	+0.66	+0.66	+0.49	+0.39	+0.46	−0.12
	Girls.		+0.33	+0.75	−0.20	+0.65	+0.43	+0.22	+0.53
Girth of Chest, Inspiration.	Boys.	Centimetre..	+0.20	+0.30	+0.17	+0.49	+0.44	+0.47	+0.41
	Girls.		+0.58	+0.28	+0.57	+0.56	+0.49	+0.55	+0.19
Girth of Chest, Expiration.	Boys.	Centimetre..	+0.34	+0.24	+0.56	+0.37	+0.22	+0.40	+0.40
	Girls.		+0.48	+0.51	+0.48	+0.48	+0.40	+0.48	+0.12
Strength of Squeeze, Right Hand......	Boys.	Kilogramme	+0.21	−0.02	+0.21	−0.70	−0.11	−0.08	−0.14
	Girls.		+0.79	+0.03	+0.13	−0.09	−0.06	+0.27	−0.07
Strength of Squeeze, Left Hand........	Boys.	Kilogramme	+0.53	−0.04	−0.02	−0.19	−0.13	−0.23	−0.29
	Girls.		+0.69	+0.54	−0.22	−0.18	−0.59	+0.02	+0.05
Length of Head.....	Boys.	Millimetre..	+0.47	+0.38	+0.82	+0.28	+0.06	+0.49	+0.24
	Girls.		+0.89	+0.49	+1.23	+1.41	+1.28	+1.67	+1.07
Width of Head.....	Boys.	Millimetre..	+0.47	+0.61	+0.33	+0.21	+0.52	+0.31	+0.38
	Girls.		+0.56	+0.70	+0.53	+0.54	+0.79	+0.49	+0.67
Root of Nose to Point of Chin.	Boys.	Millimetre..	−0.52	−0.30	−0.20	−0.09	−0.33	−0.14	+0.20
	Girls.		−0.53	−0.57	−0.05	−0.25	−0.37	−0.24	+0.07
Width of Face......	Boys.	Millimetre..	+0.96	+0.62	+0.93	+0.84	+0.03	+0.49	+0.78
	Girls.		+0.68	+0.27	−0.18	+0.79	+0.93	+0.47	+0.63
Hair Line to Point of Chin.	Boys.	Millimetre..	+0.34	+0.76	+0.54	+0.82	+0.33	+0.72	+0.39
	Girls.		+0.39	+0.83	+0.41	+0.32	+0.52	+0.85	+0.66

TABLE No. 11 — *Continued.*
MEDIAN MINUS AVERAGE VALUES.

Age at Nearest Birthday, and Median Minus Average Values.								Dimensions.
13	14	15	16	17	18	19	20	21	
—0.36 —0.49	—0.46 +0.36	—0.73 +0.16	+0.23 —0.29	—0.36 —0.09	+0.74	+0.28	—0.34		Weight.
+0.38 —0.34	+0.28 +1.10	+0.35 +0.78	+1.00 +0.51	+0.87 +0.07	+0.09 +0.08	+1.10	+0.42	+0.52	Height Standing.
+0.29 +0.49	—0.13 +0.93	—0.07 +0.78	+0.94 +0.74	+0.94 +0.75	—0.53 +0.69	+0.82	+0.69	+0.74	Height Sitting.
+0.24 +0.64	+0.34 +0.98	+0.36 —0.17	+1.35 +0.80	+1.02 +0.61	+0.19 +0.91	+0.45	+1.33	—0.39	Span of Arms.
+0.39 +0.30	+0.23 +0.43	+0.24 +0.12	+1.01 +0.57	+0.90 +0.41	+1.23 +0.48	+0.90	+0.80	—0.64	Girth of Chest, Inspiration.
+0.41 +0.40	+0.08 +0.48	+0.52 +0.17	+0.35 +0.16	+0.72 +0.26	+0.44 +0.45	+0.51	+0.41	+0.61	Girth of Chest, Expiration.
—0.09 —0.62	—0.50 —0.24	—0.52 +0.31	—1.14 —0.18	+0.37	+0.75	+0.43	—0.11		Strength of Squeeze, Right Hand.
—0.36 —0.12	—0.50 —0.69	+0.08 +0.41	+0.54 —0.08	+0.60 —0.10	+0.02	+0.38	+0.48	—0.28	Strength of Squeeze, Left Hand.
+0.37 +1.04	—0.69 +1.30	+0.68 +0.55	+0.98 +0.86	+1.48 +0.46	—0.16 —0.44	+0.40	+0.27		Length of Head.
+0.27 +0.37	+0.54 +0.35	+0.78 +0.42	+0.37 +0.72	+0.41 +0.39	+0.34 +1.07	+1.06	—0.12		Width of Head.
+0.10 —0.21	+0.08 —0.15	—0.11 —0.39	+0.10 —0.61	+0.94 +0.46	+0.16 +0.27	—0.38	+0.04	—0.06	Root of Nose to Point of Chin.
+0.19 +0.71	+0.51 +0.84	+0.71 +0.45	+0.55 +0.71	+1.79 +0.75	—0.75 +0.55	+0.20	+1.29		Width of Face.
+0.54 +0.80	+0.55 +0.62	+0.78 +0.65	—0.26 +0.53	—0.03 +0.90	+1.99 +0.96	—1.83	—0.09	+1.78	Hair Line to Point of Chin.

either may be used for the type without any error of practical importance, provided the series are similar to those analyzed here. This point is of practical interest because the labor of reckoning the average is much greater than in reckoning the median value.

Bowditch, in *The Growth of Children*, Boston, 1891, p. 495 *et seq.*, discusses the relation of median and average value. "It is evident," he writes, "that the value M will tend to "approximate to the average value of all the observations and "will be identical with it when the [percentile] curve S T "is symmetrically disposed on both sides of M, *i. e.*, when "the values at sixty, seventy, eighty, ninety and ninety-five "per cent. exceed M by the same amount, respectively, by "which the values at forty, thirty, twenty, ten and five per "cent. fall short of it. If A represent the average value of "all the observations, then the value of M–A will be a meas-"ure of the direction and extent of the asymmetry of the "curve S T, *for this value will be zero when the curve is sym-"metrical, positive when the values at the lower percentile "grades fall short of M more than those at the higher grades "exceed it, and negative when the reverse is the case.*" [Dr. Bowditch now gives a table of median minus average height and weight.] "An examination of this table or of the curves "constructed from it, as given in Plate I, shows that the "asymmetry of the curves of percentile grades varies very "much, at different ages, both in direction and amount. The "variation in the value of M–A in the curves of height is much "the same as that in the curves of weight for each sex consid-"ered by itself, but there is a great difference between the "two sexes. This difference shows itself most distinctly "between the ages of eleven and fifteen years. During this "time a rise in the curves for the males coincides with a fall "in those for the females, while before and after this period "the curves, as a rule, rise and fall together. We must con-"clude, therefore, that the rate of annual increase, both in "height and weight, is different at different percentile grades, "or, in other words, that large children grow differently from "small ones, and moreover, that between the ages of eleven "and fifteen years there is a striking difference in the mode "of growth of the two sexes."

Table No. 11 and the curves constructed from it (Plates II, III, IV) furnish considerable material for the study of the relation between median and average values. The curves of median minus average height agree with those of Dr. Bowditch in showing (1) a difference in the rate of growth of the same sex at different percentile grades, (2) an agreement in the rate of growth of the sexes from age 6 to 12, inclusive,* and (3) a marked difference in the rate of growth from age 12 to age 16, beyond which the number of observations is perhaps too small for sure work. In the curves from both cities, moreover, the greatest asymmetry in girls is near age 13, and in boys near age 16, the culmination being a little later in St. Louis children, and the least asymmetry is near age 12. The asymmetry of Boston boys' curves at age 15 and girls' curves at ages 16 and 17 is somewhat greater than that of the St. Louis curves. It is further worthy of notice that the values in the case of St. Louis boys do not become negative.

There is a general similarity also between the median minus average weights of the children of the two cities, the asymmetry of girls being again greatest about age 14 and of boys greatest at about age 16. Dr. Bowditch's conclusions, quoted above, are therefore fully confirmed.

The St. Louis curves on Plates II, III and IV should now be compared with each other. The maximum asymmetry of girls about 14 years and boys about 16 years old, appears in weight, height, height sitting, and span of arms, while the remaining curves exhibit no characteristic sexual differences during the period of prepubertal acceleration. The asymmetry is for the most part positive, weight, strength of squeeze and height of face from root of nose to point of chin being the only considerable exceptions. Sexual differences, aside from those already mentioned, are unimportant, save perhaps in length and width of head, in both of which the asymmetry of girls is greater than that of boys from age 6 to age 13.

The errors in the median minus average values of height, weight and other single dimensions may be partially compen-

---

* It should be remembered that Dr. Bowditch's ages are recorded at last birthday, while mine are recorded at nearest birthday.

sated by adding the median minus average values of several dimensions of each sex and age, without regard to plus or minus sign. This has been done in Table No. 12.

TABLE No. 12.

SUMS OF MEDIAN MINUS AVERAGE VALUES.

Age at Nearest Birthday.	Weight, Height Standing, Height Sitting, Span of Arms.		Girth of Chest at Full Inspiration and Expiration.		Strength of Squeeze, Right and Left Hand.		Head and Face Measurements.	
	Boys.	Girls.	Boys.	Girls.	Boys.	Girls.	Boys.	Girls.
6	1.55	1.48	0.54	1.06	0.74	1.48	2.76	3.05
7	1.22	1.89	0.54	0.79	0.06	0.57	2.67	2.86
8	1.86	1.60	0.73	1.05	0.23	0.35	2.82	2.40
9	1.52	1.16	0.86	1.04	0.89	0.27	2.24	3.31
10	1.51	1.58	0.66	0.89	0.24	0.65	1.27	3.89
11	1.56	1.57	0.87	1.03	0.31	0.29	2.15	3.72
12	0.71	1.92	0.81	0.31	0.43	0.12	1.99	3.10
13	1.27	1.96	0.80	0.70	0.45	0.74	1.47	3.13
14	1.21	3.37	0.31	0.91	1.00	0.93	2.37	3.26
15	1.51	1.89	0.76	0.29	0.60	0.72	3.06	2.46
16	3.52	2.34	1.36	0.73	1.08	0.26	2.26	3.43
17	3.19	1.52	1.62	0.67	....	0.47	4.65	2.96
18	....	2.42	1.77	0.93	....	0.77	3.40	3.29

The conclusions drawn from the curves discussed above are substantiated in the main by this table. It is, however, difficult to believe that a sexual difference such as exists in weight, height standing, height sitting and span of arms will not be found also in girth of chest and the various dimensions of the head. But a purely objective attitude towards the material in hand does not permit speculation as to what might be revealed if the material were very much larger and its intrinsic laws easier to unveil.

CHAPTER IV.

DANGERS OF COMPARING MEASUREMENTS OF CHILDREN IN DIFFERENT COUNTRIES. INFLUENCE OF OCCUPATION AND NATIONALITY OF PARENTS.

It has been a custom of anthropometrists to compare the type-children of cities, states or provinces very different in situation and character. Such comparisons may be expected always to show that the laws of growth are in their main features the same for all children. They cannot, however, be expected to give very definite information in regard to the relative size of children of different countries or districts so long as the social status or environment of the children is not more closely studied. The children in the public schools are from all classes of society, and it has been demonstrated repeatedly that the favored classes differ physically from the poor. The children of the prosperous have been found to be larger than the children of the poor. The comparison of middle values got from two sets of schools is therefore open to the objection that the composition of the school population may not be the same in both sets. It would for example be unsafe to say that St. Louis children are larger or smaller than Copenhagen children because the type-children in the St. Louis Public Schools are larger or smaller than the type-children in Copenhagen, for the difference observed may depend on the different composition of the school population in the two cities.

Such comparisons, when rightly made, are not only of great scientific interest but are almost essential to the use of anthropometrical systems in education. An immense saving of time would be made if it were shown that the typical height, weight, etc., of children in one city of a country could be adopted as the standard for the schools of the entire country or even a considerable part of it. And in any one city, the application of the type-values to individuals would be much less liable to error if it were known how much allowance should be made for the difference between the type of a special social class and the types of a mixed school population. Every investigation of the present kind should, therefore, include information concerning the social condition of the children. The tax-returns are available for this purpose, and the occupation of parents is also of use. It was not possible to inquire into tax-returns in the present research, but some facts can be communicated regarding the physical differences between the children of professional and business men and those of manual tradesmen. The occupations included under these heads may be seen in the following statement of the occupations of the fathers of 2,000 St. Louis children. The names of occupations are set down as given by the pupils.

PROFESSIONS.

Artist...............................	3	Minister........................	6
Chemist...........................	1	Medical Student..............	1
Civil Engineer...................	3	Musician	6
Dentist.............................	4	Photographer..................	3
Editor.............................	6	Physician.......................	13
Electrical Engineer............	4	Teacher.........................	9
Electrician.......................	2	Veterinarian....................	1
Lawyer............................	14		
Total.........................			76

MERCANTILE CLASS.

Agent...............................	36	Dealer in Furniture...........	6
Banker.............................	2	Hardware................	2
Book-keeper.....................	40	Hats........................	1
Broker.............................	9	Horses.....................	1
Cashier............................	8	Lime........................	1
Clerk...............................	3	Lumber...................	4
Collector..........................	8	Marble....................	1
Contractor.......................	20	Pork.......................	2
Dairyman.........................	8	Produce	2
Drugs..............................	10	Stationery..............	1
Dealer in Butter.................	1	Stone.....................	1
Carpets.......................	2	Tobacco.................	5
Cattle..........................	4	Wall-paper.............	1
Cloth...........................	4	Foreman	24
Clothes........................	1	Grocer	36
Corn............................	2	Hotel-keeper...............	5
Coal............................	4	Inspector, Building	2
Feed and Flour............	2	Street	7
Fish and Oysters..........	3	Not specified........	1
Fruits.........................	3	Jeweler........................	7

Insurance	4	President of Iron Co	1
Manager	7	Publisher	1
Manufacturer	35	Real Estate	7
Merchant	143	Salesman	76
Merchant Tailor	2	Secretary of a corporation	6
News-depot	1	Telegrapher	1
Optician	2	Traveling Salesman	9
Postmaster	1	Undertaker	3
Total			579

MANUAL TRADES.

Baker	24	Maker of:	
Barber	23	Boilers	7
Blacksmith	29	Boxes	9
Book-Binder	2	Brushes	3
Brewer	14	Candies	3
Bricklayer	21	Carriages	16
Builder	9	Chains	2
Butcher	40	Chairs	2
Carpenter	148	Cigars	31
Carriage Repairer	2	Collars	1
Cigar Packer	1	Cornices	1
Cloth Sponger	1	Harness	2
Compositor	2	Lanterns	1
Confectioner	7	Mattresses	1
Cook	5	Models	1
Cooper	37	Organs	1
Coppersmith	1	Patterns	4
Cord-Wainer	1	Shoes	47
Dairyman	1	Tobacco	3
Decorator	4	Trunks	1
Draughtsman	5	Wagons	6
Dressmaker	4	Watches	6
Dyer	2	Machinist	55
Engineer	43	Maltster	4
Engraver	1	Marble Cutter	3
Fireman	11	Mechanic	42
Foundryman	2	Miller	7
Gardener	5	Millwright	5
Gilder	1	Miner	1
Glass Blower	2	Molder	33
Iron Worker	3	Motorman	2
Laundryman	1	Packer	5
Lithographer	1	Painter	51
Locksmith	1	Paper-Hanger	8
Maker of:		Paver	4
Bags	2	Pipe-Fitter	2
Baskets	1	Planer	1
Belts	1	Plasterer	10

Plumber	13	Tailor	56
Polisher	1	Tanner	6
Potter	2	Tinner	15
Printer	19	Turner	6
Puddler	1	Upholsterer	5
Saddler	12	Varnisher	5
Selector of Tobacco	1	Waterproofer and Painter	1
Shoe Varnisher	1	Weaver	1
Slater	1	Whitener	14
Stair Builder	4	Wood-Worker	33
Stone Mason	50		
Total			1086

LABORERS.

Baggageman	2	Oiler	2
Carriers, Letter	9	Peddler	19
Carriers, Paper	6	Policeman and watchman	21
Coal-heaver	1	Porter	4
Driver	3	Riverman	1
Elevator man	1	Railway-hand	12
Engine-cleaner	1	Saloon-keeper	51
Expressman	3	Sexton	1
Fireman	11	Shepherd	1
Gripman	6	Steward, Hotel	2
Hostler	7	" Steamer	1
Huckster	2	"	3
Janitor	3	Teamster	6
Laborer	31	Waiter	1
Live-stock driver	1		
Lumberman	1	Total	216
Mill-hand	4		

MISCELLANEOUS.

Captain of Steamer	1
Conductor	14
Farmer	12
"Independent"	1
Pilot	5
No occupation	7
"Retired"	3
Total	43

SUMMARY.

Professions	76
Mercantile classes	579
Manual trades	1086
Laborers	216
Miscellaneous	43
Total	2000

The number of children of rich families and of very poor families in the public schools is small, the larger part of the pupils being from what would be called in England the lower middle class, and the school population is accordingly more homogeneous than would have seemed possible at the first glance.

The difference in the weight of children of the professional and merchant class is probably of little importance until the period of prepubertal acceleration. Such at least is the inference from Table No. 13, where the weights of these two classes are compared.

TABLE No. 13.

A COMPARISON OF THE WEIGHTS OF THE DAUGHTERS OF MANUAL TRADESMEN WITH THE WEIGHTS OF THE DAUGHTERS OF PROFESSIONAL MEN AND MERCHANTS.

Professional Men and Merchants.			Manual Tradesmen.		
Age at Nearest Birthday.	Number of Observations.	Median Weight. (Kilogrammes)	Median Weight. (Kilogrammes)	Number of Observations.	Age at Nearest Birthday.
6	74	19.53	18.95	237	6
7	148	20.63	20.74	604	7
8	170	23.00	22.89	723	8
9	152	25.48	25.09	688	9
10	168	27.61	27.46	651	10
11	173	30.63	29.45	569	11
12	153	33.86	32.28	556	12
13	160	39.04	37.07	402	13
14	140	43.59	41.68	251	14
15	112	47.49	45.62	145	15
16	87	50.16	49.18	52	16
17	46	53.58	51.07	24	17

What is true of weight is in this instance probably true of the physical development as a whole, and Table No. 13 would seem to indicate that a wide difference in social status or material prosperity may exist without much influencing the growth of children up to the prepubertal acceleration. But further investigation is necessary before a generalization can be made. It must suffice here to state: 1. The daughters of professional men and merchants are very little heavier than the daughters of manual tradesmen until the period of prepubertal acceleration. 2. The weight of girls is much more influenced by the material prosperity or social status of parents during and immediately after the period of pre-pubertal acceleration than in the earlier years of growth.

The proportion of children of the more prosperous class in the public schools is not the same in all grades. In the higher grades, containing the older children, the number from prosperous families is relatively larger, the children of the poor having in many instances been compelled to leave school in order to earn money. Some idea of the extent of this change may be gained from Table No. 14, in which the per cent. of daughters of professional men and merchants and daughters of manual tradesmen is given at each age and school grade.

TABLE No. 14.

THE PERCENTILE DISTRIBUTION BY SCHOOL GRADE OF THE DAUGHTERS OF MERCHANTS AND PROFESSIONAL MEN (*i. e.* FAVORED CLASSES) COMPARED WITH THAT OF THE DAUGHTERS OF MANUAL TRADESMEN (HAND-WORKERS).

Age at nearest Birthday.	Number of Individuals.	Percentile Distribution by School Grades.								
		Kindergarten.	I.	II.	III.	IV.	V.	VI.	VII.	VIII.	
Six............	44 141	95.4 91.5	4.6 8.5								Professions. Trades.
Seven..........	102 341	46.1 40.0	48.0 54.8	5.9 5.2							Professions. Trades.
Eight..........	105 440	2.0 2.0	49.5 65.4	46.6 30.9	2.0 1.4	0.2					Professions. Trades.
Nine...........	110 439		10.0 23.7	60.0 57.8	23.6 16.2	4.5 2.3	1.8				Professions. Trades.
Ten............	119 386		1.7 5.7	26.9 39.9	45.4 44.3	23.5 8.5	1.7 1.0	0.8 0.5			Professions. Trades.
Eleven.........	116 355		0.9 2.5	12.0 15.5	24.1 41.2	50.0 33.0	11.2 7.6	0.9 0.8			Professions. Trades.
Twelve.........	108 357		0.9 0.6	3.7 7.6	13.9 19.6	37.0 40.0	28.7 21.8	13.0 9.8	0.9 0.3	1.9	Professions. Trades.
Thirteen........	121 304		0.7	2.5 3.3	4.1 12.2	18.9 27.3	31.4 25.0	24.8 20.7	11.6 9.9	6.6 1.0	Professions. Trades.
Fourteen.......	89 177			1.1	3.4 4.5	5.6 14.7	18.0 25.4	32.6 29.2	20.2 15.8	20.2 9.0	Professions. Trades.
Fifteen.........	65 110				1.5 1.8	1.5 7.1	10.8 14.5	29.2 32.7	21.5 20.0	35.4 23.6	Professions. Trades.
Sixteen.........	37 24					2.7 4.2	2.7 8.3	21.6 16.7	35.1 29.2	37.7 41.7	Professions. Trades.

The ratio between the two is more easily grasped when the difference between the classes at each age and grade is stated, black type being used where the professional and merchant class exceed the tradesmen's daughters and lower case type where the contrary is true.

Age.	Kinder-garten.	I.	II.	III.	IV.	V.	VI.	VII.	VIII.
6	**3.9%**	3.9%	%	%	%	%	%	%	%
7	**6.1**	6.8	0.7						
8	0.0	15.9	15.7	0.6					
9		13.7	2.2	7.4	2.2				
10		4.0	13.0	1.1	15.0	0.7	0.3		
11		1.6	3.5	17.1	17.0	3.6	0.1		
12		**0.3**	3.9	5.7	3.0	6.9	3.2	0.6	
13			0.8	8.1	8.4	6.4	**4.1**	1.7	5.6
14				1.1	9.1	7.4	3.4	4.4	11.2
15				0.3	5.6	3.7	3.5	**1.5**	11.8
16					1.5	5.6	**4.9**	5.9	4.0

The number of boys withdrawn from school to earn money is of course greater than the number of girls withdrawn. Thus of 562 boys, aged 6, 17.8 per cent. were sons of professional men or merchants and 42.9 per cent. sons of manual tradesmen, while at age 14, 29.3 per cent. of 498 boys were from the former class and 37 per cent. from the latter.

Age at Nearest Birthday.	Number of Individuals.	Sons of Professional Men and Merchants.	Sons of Manual Tradesmen.	Others.
6	562	17.8%	42.9%	39.3%
14	498	29.3	37.0	33.7

The nationality of the children should be considered in an anthropometrical inquiry. It is well known that children of the same age but different nationality exhibit differences in physical development. The annual report of the Superintendent of Public Schools, issued Aug. 1, 1891, contains in Table V, appendix, page XLVI, a statement "Showing the Birthplace of Pupils Registered in Each School for the Year 1890–91," from which the number of pupils in the schools in which the measurements were made has been taken and the following percentages calculated: —

BIRTHPLACE OF 46,870 PUPILS IN THE ST. LOUIS PUBLIC SCHOOLS.

St. Louis	79.26 %
Other Parts of the United States	16.92 "
Great Britain	0.63 "
Ireland	0.19 "
German States	1.97 "
Other Foreign Countries	0.87 "
Unknown	0.16 "
Total	100

The children of foreign birth are too few to affect the results of the measurements, and the number of children born in or near St. Louis is such that the middle values obtained must be taken as characteristic for this community.

The children of foreign parentage are of course much more numerous than those of foreign birth. The number of Germans is especially large. The median weights of boys and girls whose parents were born in Germany are compared in Tables No. 15 and 16, with children of American parentage.

TABLE No. 15.*

THE WEIGHTS OF GIRLS WHOSE PARENTS WERE BORN IN GERMANY COMPARED WITH THE WEIGHTS OF GIRLS WHOSE PARENTS WERE BORN IN THE UNITED STATES.†

GERMAN.			AMERICAN.		
Age at Nearest Birthday.	Number of Observations.	Median Weight. (Kilogrammes)	Median Weight. (Kilogrammes)	Number of Observations.	Age at Nearest Birthday.
6	310	19.15	18.76	398	6
7	683	20.86	20.82	861	7
8	796	23.17	22.71	1082	8
9	796	25.09	25.07	1023	9
10	725	27.65	27.43	1029	10
11	753	29.61	29.93	808	11
12	715	33.42	33.17	779	12
13	518	37.58	38.29	648	13
14	331	42.56	43.12	565	14
15	186	46.77	46.90	403	15
16	86	49.73	50.06	265	16
17	49	53.93	52.12	131	17
18	43	52.59	54.03	100	18
19	22	54.26	52.90	40	19
20			52.67	33	20

* The birth-place of the girls themselves is not considered in this table.
† A few were born in Canada.

TABLE No. 16.*

THE WEIGHTS OF BOYS WHOSE PARENTS WERE BORN IN GERMANY COMPARED WITH THE WEIGHTS OF BOYS WHOSE PARENTS WERE BORN IN THE UNITED STATES.†

GERMAN.			AMERICAN.		
Age at Nearest Birthday.	Number of Observations.	Median Weight. (Kilogrammes)	Median Weight. (Kilogrammes)	Number of Observations.	Age at Nearest Birthday.
6	158	20.04	19.66	263	6
7	334	21.93	21.67	756	7
8	426	23.98	23.91	907	8
9	369	26.64	26.08	878	9
10	370	28.51	28.49	847	10
11	358	31.21	31.26	663	11
12	385	33.51	33.45	549	12
13	321	35.92	35.96	437	13
14	166	39.59	40.34	352	14
15	106	44.68	47.25	219	15
16	26	52.22	52.10	92	16
			55.14	40	17

The difference in weight is seen to be of no great importance.

In the absence of special investigations of the influence of social condition and the nationality of parents on the growth of children, it is impossible to estimate accurately to what degree middle values, calculated without regard to social condition and nationality, are affected by these factors. The data presented in this chapter go to show that the middle values of St. Louis children are little influenced by considerable differences in social condition before the period of prepubertal acceleration and are not very largely influenced by such differences during this period. Further, these values are at no time much affected by differences in nationality of parents not greater than those existing between Germans and Americans. But the whole question evidently requires extended study of data difficult or impossible of collection by private hands.

---

* The birth-place of the boys themselves is not considered in this table.
† A few were born in Canada.

CHAPTER V.

PERCENTILE GRADES OF WEIGHT, HEIGHT STANDING, HEIGHT SITTING, SPAN OF ARMS, GIRTH OF CHEST, LENGTH OF HEAD, WIDTH OF HEAD, HEIGHT OF FACE FROM ROOT OF NOSE TO POINT OF CHIN, WIDTH OF FACE, AND HEIGHT OF FACE FROM HAIR LINE TO POINT OF CHIN.

The Percentile Grades of weight, height, etc., presented in Tables No. 17 to No. 28, inclusive, are represented graphically in Plates V to XXIV. By their aid, the percentile rank of an individual in respect of weight, span of arms or any physical dimension included in the tables can be easily and quickly determined. Suppose, for example, the percentile rank of a boy, aged 11, weighing 32 kg. was desired. A horizontal line is drawn from 32 in the column of kilogrammes on the left of Plate V to the curve of age 11, and a perpendicular is dropped from the point of intersection to the scale of percentile grades at the bottom of the plate. The perpendicular falls at 60 per cent. Hence the boy is heavier than 60 per cent. of boys of his age and lighter than 40 per cent.

Other facts are made plain by these curves. The increase at any percentile grade during one or more years is expressed by the distance between the curves at that grade. Thus, in Plate V, the gain in weight of the type-boy in the 15 percentile grade during the years 6 to 11 inclusive is 9.3 kg., and the gain at the 85 percentile grade during the same year 13.0 kg. The difference in size between large and small boys of the same age appears in the inclination of the curve to its axis, the slope being steepest in the years of quickest growth. And finally, the tendency of the greater number of individuals to approximate the middle value of their group is illustrated, the inclination of each curve being much less at the central part than at the ends, where the giants and the dwarfs are found. The principal service of such curves, however, is to determine percentile rank.

TABLE No. 17.
The Weight.

Age at Nearest Birthday.	Sex.	Number of Observations.	Value in Kilogrammes at the following Percentile Grades.										Average.
			5	10	20	30	40	50	60	70	80	90	95	Kilogram.
Six	Boys.	707	16.51	17.37	18.31	18.83	19.35	19.85	20.32	20.82	21.71	22.81	23.78	19.75
	Girls.	796	16.00	16.60	17.39	18.01	18.45	18.99	19.49	20.03	20.85	22.06	22.99	18.93
Seven	Boys.	1814	18.17	18.85	19.77	20.50	21.03	21.66	22.24	22.91	23.80	25.26	26.50	21.67
	Girls.	1714	17.16	17.95	18.88	19.62	20.24	20.81	21.39	22.13	22.96	24.36	25.75	20.82
Eight	Boys.	2188	19.49	20.44	21.60	22.40	23.10	23.87	24.60	25.36	26.34	27.89	29.15	23.78
	Girls.	2147	18.83	19.61	20.64	21.40	22.17	22.85	23.60	24.31	25.29	27.08	28.38	22.88
Nine	Boys.	2188	21.56	22.65	23.78	24.59	25.30	26.22	26.93	27.78	28.77	30.15	32.19	26.06
	Girls.	2055	20.45	21.40	22.65	23.49	24.25	25.04	25.82	26.77	27.85	29.60	31.24	25.08
Ten	Boys.	2064	22.97	24.19	25.67	26.57	27.53	28.36	29.34	30.19	31.42	33.18	34.76	28.32
	Girls.	1947	21.93	23.18	24.49	25.66	26.67	27.45	28.28	29.33	30.58	32.29	34.16	27.49
Eleven	Boys.	1743	25.31	26.35	27.89	29.11	30.03	31.08	32.01	32.96	34.35	36.14	37.97	31.00
	Girls.	1708	23.38	25.26	26.78	27.78	28.86	29.80	30.94	32.12	33.56	36.37	38.35	30.15
Twelve	Boys.	1644	27.47	28.57	30.05	31.25	32.43	33.41	34.39	35.59	36.93	39.21	41.43	33.51
	Girls.	1676	26.23	27.70	29.40	30.70	31.88	33.25	34.50	35.98	37.67	40.98	44.78	33.66
Thirteen	Boys.	1242	29.29	30.47	32.14	33.62	34.82	36.25	37.37	38.90	40.81	43.40	47.38	36.61
	Girls.	1343	29.19	30.78	33.05	34.67	36.25	38.00	39.81	41.80	43.77	47.49	50.25	38.49
Fourteen	Boys.	946	32.02	33.67	35.70	37.11	38.50	39.98	41.98	42.98	46.78	50.32	53.96	40.44
	Girls.	1082	31.50	34.14	36.73	39.01	40.87	42.65	44.30	46.48	48.61	51.94	54.99	42.29
Fifteen	Boys.	498	35.76	37.05	39.02	41.05	43.17	45.49	47.76	50.17	53.33	57.11	60.29	46.22
	Girls.	690	36.48	38.54	41.81	43.37	45.26	46.85	48.05	49.65	51.67	54.86	58.05	46.69
Sixteen	Boys.	203	37.72	39.87	44.55	46.77	49.51	51.83	54.53	56.88	58.99	63.87	67.16	51.60
	Girls.	420	39.86	42.09	45.63	47.49	48.87	49.96	51.56	53.08	55.84	59.31	61.56	50.25
Seventeen	Boys.	71	45.45	47.45	51.34	52.54	54.30	55.31	57.68	60.24	62.17	63.91	64.87	55.67
	Girls.	230	44.27	45.66	47.78	49.56	51.30	52.52	54.28	55.61	57.32	60.15	63.22	52.61
Eighteen	Girls.	155	45.65	46.99	48.51	50.03	52.06	53.10	54.23	55.70	57.96	60.76	64.33	52.36
Nineteen	Girls.	81	44.05	45.45	47.87	49.99	51.48	52.47	53.62	54.68	56.80	59.84	61.72	52.19
Twenty	Girls.	66	45.54	47.31	49.08	50.94	52.72	53.57	55.02	55.57	59.09	63.02	66.11	53.91

Porter — *The Growth of St. Louis Children.* 313

TABLE No. 18.
THE HEIGHT STANDING.

Age at Nearest Birthday.	Sex.	Number of Observations.	Value in Centimetres at the following Percentile Grades.										Average.
			5	10	20	30	40	50	60	70	80	90	95	Centimeters
Six	Boys.	709	101.77	103.50	105.25	106.73	107.90	109.23	110.40	111.86	113.51	115.81	118.32	108.94
	Girls.	780	100.20	101.41	103.97	105.51	106.78	108.10	109.40	110.61	111.98	115.00	116.90	107.67
Seven	Boys.	1850	105.67	107.56	110.23	111.81	112.97	114.48	115.81	117.13	118.90	121.24	123.39	114.03
	Girls.	1791	104.93	106.24	108.93	110.51	111.89	113.44	115.03	116.32	117.82	120.38	121.94	112.95
Eight	Boys.	2225	109.68	112.28	115.15	117.01	118.44	119.78	121.23	122.72	124.44	126.73	129.05	119.13
	Girls.	2193	110.18	111.86	114.32	115.82	117.40	118.75	120.10	121.42	123.27	125.79	127.79	118.36
Nine	Boys.	2205	115.89	118.11	120.29	121.99	123.38	124.87	126.25	127.87	129.64	131.90	134.06	124.35
	Girls.	2122	115.17	117.12	119.59	121.21	122.76	124.11	125.53	126.64	128.62	130.97	133.42	123.67
Ten	Boys.	2087	120.04	121.92	124.58	126.40	127.98	129.45	130.98	132.54	134.55	137.14	139.45	128.87
	Girls.	2053	119.43	121.34	124.14	125.77	127.32	128.85	130.33	131.82	133.33	136.86	138.84	128.43
Eleven	Boys.	1819	124.53	126.59	129.11	130.95	132.75	134.44	135.85	137.24	139.33	142.18	144.75	133.84
	Girls.	1772	122.95	125.41	128.16	130.29	132.08	133.60	135.25	136.87	139.06	142.16	144.69	133.19
Twelve	Boys.	1655	128.65	130.59	133.26	135.23	136.87	138.57	140.41	141.94	144.05	147.02	149.83	139.21
	Girls.	1732	128.06	130.66	133.63	135.68	137.55	139.54	141.38	143.29	145.68	149.16	152.19	139.11
Thirteen	Boys.	1268	131.36	134.14	137.39	139.51	141.39	143.29	145.12	147.01	149.54	153.55	155.91	142.91
	Girls.	1322	134.16	137.09	139.84	142.12	144.23	146.19	148.26	150.08	152.90	156.03	158.63	146.53
Fourteen	Boys.	925	137.10	139.63	142.30	144.56	146.50	148.86	150.67	152.26	155.77	159.19	162.88	148.68
	Girls.	1085	139.02	141.66	144.68	148.01	150.25	151.94	153.70	155.37	157.26	160.31	161.99	150.84
Fifteen	Boys.	490	140.94	143.69	147.50	150.26	152.65	155.25	157.47	160.19	163.53	168.00	170.44	154.90
	Girls.	680	145.11	148.20	150.82	152.63	154.28	155.82	156.89	158.40	160.65	162.77	164.83	155.04
Sixteen	Boys.	189	146.42	149.38	153.93	156.62	159.45	161.27	163.08	165.47	168.60	171.02	173.55	160.27
	Girls.	420	149.00	150.45	153.05	154.82	156.52	158.03	159.54	160.97	162.65	165.93	167.64	157.52
Seventeen	Boys.	78	154.90	156.90	159.72	161.85	164.20	166.00	168.13	170.20	172.48	175.07	177.05	165.13
	Girls.	206	150.43	152.46	154.73	156.57	157.77	159.40	160.54	162.02	163.83	166.23	168.46	159.33
								170.50						
Eighteen	Girls.	164	151.10	153.47	155.73	157.13	158.54	159.50	160.57	162.26	164.31	166.45	169.60	159.42
Nineteen	Girls.	85	150.13	151.36	153.00	155.64	158.43	159.56	160.83	162.42	164.11	165.17	167.76	158.46

TABLE No. 19.
THE HEIGHT SITTING.

Age at Nearest Birthday.	Sex.	Number of Observations.	Value in Centimetres at the following Percentile Grades.											Average. Centimetres
			5	10	20	30	40	50	60	70	80	90	95	
Six.......	Boys.	714	56.28	57.50	58.71	59.50	60.17	60.77	61.52	62.42	63.61	65.31	68.58	61.31
	Girls.	751	55.19	56.28	57.68	58.78	59.41	60.11	60.80	61.59	62.50	63.68	64.75	59.46
Seven.....	Boys.	1853	58.41	59.50	60.72	61.60	62.42	63.22	63.98	64.77	65.66	67.38	69.19	63.32
	Girls.	1727	57.41	58.65	60.16	60.99	61.80	62.44	63.28	64.12	65.00	66.08	67.20	61.80
Eight......	Boys.	2239	60.21	61.38	62.70	63.74	64.52	65.21	65.87	66.69	67.65	69.01	70.32	64.74
	Girls.	2120	60.00	61.00	62.27	63.14	64.01	64.96	65.22	65.82	66.92	68.21	69.27	63.97
Nine.......	Boys.	2258	62.10	63.16	64.59	65.50	66.29	67.08	67.93	68.82	69.79	71.17	72.43	66.73
	Girls.	2071	61.79	62.99	64.26	65.30	66.00	66.19	67.52	68.33	69.28	70.59	71.81	66.16
Ten.......	Boys.	2118	63.70	65.49	66.35	67.45	68.37	69.18	69.98	70.84	71.86	73.46	74.79	69.25
	Girls.	2037	63.46	64.86	66.19	67.22	68.06	68.88	69.61	70.48	71.49	72.79	74.02	68.19
Eleven.....	Boys.	1828	65.23	66.62	68.29	69.27	70.23	71.09	71.84	72.70	73.75	75.23	76.63	70.67
	Girls.	1748	65.15	66.27	67.83	68.08	69.89	70.62	71.42	72.36	73.51	75.08	76.37	70.03
Twelve....	Boys.	1656	67.62	68.35	70.08	70.89	71.84	72.68	73.56	74.50	75.56	77.00	78.81	72.55
	Girls.	1707	66.06	68.46	70.21	71.16	72.39	73.22	74.21	75.26	76.29	78.24	79.09	72.67
Thirteen...	Boys.	1285	68.68	70.02	71.45	72.61	73.58	74.49	75.75	76.27	77.48	79.24	81.00	74.20
	Girls.	1354	69.99	71.44	73.15	74.40	75.46	76.52	77.64	78.47	80.41	82.06	83.48	76.03
Fourteen...	Boys.	936	70.49	71.78	73.76	74.97	75.83	76.71	77.84	79.03	80.71	82.85	85.37	76.84
	Girls.	1065	71.54	73.83	75.79	77.20	78.29	79.61	80.59	81.59	82.87	84.74	85.95	78.68
Fifteen.....	Boys.	498	72.66	74.28	75.84	77.17	78.41	79.67	81.08	82.70	84.75	86.58	88.51	79.74
	Girls.	674	75.34	77.05	79.17	80.34	81.27	82.20	83.12	84.04	85.12	86.55	87.72	81.42
Sixteen....	Boys.	193	74.66	76.37	78.73	80.00	82.00	83.22	84.57	85.78	86.86	89.14	91.39	82.28
	Girls.	411	78.93	80.77	81.67	82.65	83.54	84.50	85.34	86.00	87.03	88.11	89.57	83.76
Seventeen...	Boys.	77	77.84	79.84	82.47	84.22	85.42	86.22	88.03	89.16	90.00	92.32	94.07	85.68
	Girls.	202	80.44	81.62	82.19	83.81	84.66	85.41	86.20	86.90	87.70	89.10	90.32	84.66
Eighteen...	Boys.	31		84.10	85.40	86.32	87.08	87.70	89.60	91.57	93.20	93.98	94.72	88.23
	Girls.	167	81.47	82.44	83.78	84.74	85.87	85.89	86.51	87.18	87.94	88.85	90.87	85.20
Nineteen...	Girls.	85	80.81	81.58	83.50	84.42	86.07	85.68	86.29	86.89	87.64	88.56	89.75	84.86
Twenty....	Girls.	78	80.95	82.13	83.65	84.64	85.35	86.00	86.78	87.66	88.49	89.60	90.28	85.31

The Span of Arms.

Value in Centimetres at the following Percentile Grades.

Age at Nearest Birthday.	Sex.	Number of Observations.	5	10	20	30	40	50	60	70	80	90	95	Average.
Six	Boys.	708	100.76	102.45	105.00	106.32	108.27	109.57	110.70	112.94	114.26	117.00	119.00	108.95
	Girls.	769	98.43	100.36	102.87	105.49	106.57	106.96	108.77	110.42	112.39	115.16	117.07	106.96
Seven	Boys.	1862	105.23	107.30	110.17	112.00	113.58	115.08	116.19	117.88	119.85	122.26	124.18	114.42
	Girls.	1724	103.14	105.21	107.60	109.81	111.34	113.11	114.70	116.63	117.91	120.65	122.76	112.36
Eight	Boys.	2234	110.28	112.75	115.60	117.44	119.13	120.73	122.32	123.66	125.76	128.43	130.49	120.07
	Girls.	2152	108.52	111.85	113.79	115.61	117.29	118.13	120.38	121.91	123.11	126.80	128.90	118.34
Nine	Boys.	2272	115.76	118.18	120.70	122.40	124.14	125.67	127.15	128.99	130.86	133.84	136.46	125.18
	Girls.	2065	113.60	115.85	119.19	121.11	122.57	124.28	125.75	127.37	129.46	131.96	134.75	123.63
Ten	Boys.	2076	119.57	122.15	125.21	127.16	129.03	130.61	132.47	134.48	136.49	139.66	141.74	130.22
	Girls.	2045	118.24	120.67	123.63	127.30	127.30	129.21	130.77	132 54	134.94	138.81	140.80	128.75
Eleven	Boys.	1810	124.46	126.77	129.92	131.87	133.79	135.59	137.58	139.64	141.50	144.67	147.26	135.13
	Girls.	1757	122.78	125.27	128.22	130.62	132.61	134.56	136.17	138.17	140.78	144.92	147.00	134.24
Twelve	Boys.	1664	128.55	131.00	134.51	136.58	138.69	140.48	142.30	144.35	146.50	150.18	153.19	140.60
	Girls.	1718	127.45	130.34	134.74	138.33	138.59	140.60	142.50	144.74	147.36	151.32	154.36	140.07
Thirteen	Boys.	1281	132.33	135.35	138.75	141.40	143.28	145.33	147.43	150.05	152.69	156.96	159.62	145.09
	Girls.	1368	134.95	137.49	140.69	143.50	145.88	147.83	150.26	152.18	154.59	158.40	160.55	147.19
Fourteen	Boys.	934	138.21	141.06	144.72	147.28	149.48	151.62	154.11	156.00	158.72	163.20	167.32	151.28
	Girls.	1088	139.60	142.92	146.81	149.79	151.84	153.56	155.36	157.38	159.70	162.38	164.70	152.58
Fifteen	Boys.	495	140.96	145.36	150.32	153.30	156.21	158.79	161.24	164.00	168.07	173.17	177.53	158.43
	Girls.	677	145.68	148.48	151.29	153.94	155.32	156.21	158.52	160.43	162.39	165.55	168.24	156.38
Sixteen	Boys.	189	145.82	150.22	156.30	159.78	162.74	165.31	167.91	169.88	173.10	176.89	180.55	163.96
	Girls.	413	148.13	150.55	154.35	156.43	157.61	159.31	160.80	162.26	164.26	167.39	170.15	158.51
Seventeen	Boys.	75	154.50	159.50	162.66	165.87	167.75	169.58	170.83	174.10	176.66	180.25	183.50	169.56
	Girls.	202	149.20	151.40	154.43	156.29	157.88	159.62	160.88	162.53	164.37	167.13	169.30	159.01
Eighteen	Boys.	32	166.20	168.10	170.40	171.87	172.93	175.50	177.40	178.47	181.40	184.80	186.40	175.31
	Girls.	164	150.30	152.68	154.54	157.76	160.05	161.38	162.65	164.02	166.30	169.10	170.93	160.47
Nineteen	Girls.	83	147.00	151.30	154.60	155.74	157.05	158.90	160.83	161.89	164.28	168.70	170.28	158.46
Twenty	Girls.	76	151.80	153.80	155.64	157.26	158.80	161.50	162.93	164.03	165.93	168.47	169.55	160.17
Twenty-one	Girls.	37	153.70	156.47	158.33	158.80	160.88	163.80	165.96	167.40	169.50	171.15	161.27	

316   Trans. Acad. Sci. of St. Louis.

TABLE No. 21.
THE GIRTH OF CHEST AT FULL INSPIRATION.

Age at Nearest Birthday.	Sex.	Number of Observations.	Value in Centimetres at the following Percentile Grades.										Average.
			5	10	20	30	40	50	60	70	80	90	95	Centimetr's
Six	Boys.	674	56.05	57.13	58.72	59.87	60.36	61.04	61.73	62.59	63.75	65.59	67.04	60.84
	Girls.	739	54.77	56.41	57.52	58.64	59.51	60.26	60.92	61.69	62.81	64.42	66.04	59.68
Seven	Boys.	1702	57.74	58.98	60.26	61.25	62.47	62.86	63.70	64.69	65.82	67.45	69.37	62.56
	Girls.	1603	55.99	57.26	58.75	59.93	60.75	61.43	62.39	63.34	64.55	66.14	67.68	61.15
Eight	Boys.	2114	59.32	60.57	62.14	63.10	64.00	64.46	65.76	66.70	67.73	69.21	70.23	64.29
	Girls.	2044	57.80	58.91	60.28	61.32	62.27	63.19	64.18	65.03	66.10	67.69	68.79	62.62
Nine	Boys.	2118	61.33	62.18	63.61	64.68	65.64	66.51	67.41	68.42	69.59	71.16	72.49	66.02
	Girls.	1964	59.87	60.33	61.95	63.10	64.06	64.87	65.73	66.80	67.89	69.52	71.00	64.31
Ten	Boys.	2000	62.33	63.68	65.20	66.60	67.40	68.31	69.23	70.23	71.46	73.41	75.30	67.87
	Girls.	1891	59.36	60.26	61.85	63.00	63.93	66.46	67.41	68.46	71.15	72.81	65.97	65.97
Eleven	Boys.	1732	64.03	65.32	67.13	68.33	69.28	70.16	70.92	71.97	73.36	75.44	77.25	69.69
	Girls.	1643	62.35	63.26	64.91	66.16	67.38	68.45	69.61	70.48	71.61	73.84	75.81	67.90
Twelve	Boys.	1568	65.44	66.86	68.66	69.93	70.91	71.84	72.97	73.99	75.32	77.30	78.96	71.43
	Girls.	1628	63.64	65.31	67.29	68.50	69.55	70.61	71.77	73.11	74.81	77.11	79.23	70.42
Thirteen	Boys.	1240	66.38	68.20	70.73	71.54	72.59	73.70	74.83	76.17	77.76	79.95	82.00	73.31
	Girls.	1315	65.39	67.53	69.85	71.20	72.37	73.65	74.93	76.34	77.96	80.62	82.84	73.35
Fourteen	Boys.	920	69.21	70.60	72.43	73.90	75.25	76.44	77.95	79.32	80.96	83.53	85.74	76.21
	Girls.	1021	68.44	70.33	72.46	73.98	75.48	76.82	78.34	79.77	81.34	83.61	85.68	76.39
Fifteen	Boys.	497	71.55	73.43	75.68	77.12	78.62	79.99	81.58	83.27	85.23	87.71	89.57	79.75
	Girls.	660	71.19	73.08	75.17	76.69	78.26	79.07	80.64	82.71	83.84	85.87	88.25	78.95
Sixteen	Boys.	206	72.10	74.65	78.02	79.75	81.90	83.60	84.97	86.65	88.02	91.06	93.17	82.59
	Girls.	398	74.45	75.85	77.70	79.28	80.58	81.65	82.70	84.57	85.78	88.20	89.72	81.08
Seventeen	Boys.	79	73.75	78.90	81.83	83.45	84.77	86.08	87.23	88.61	90.30	93.28	94.85	85.18
	Girls.	211	75.92	77.14	79.46	80.75	81.92	83.18	84.25	85.34	87.07	89.49	91.29	82.77
Eighteen	Boys.	31	78.10	84.05	86.20	88.10	89.10	89.88	90.52	91.70	92.70	93.95	95.45	88.65
	Girls.	164	76.04	77.85	79.34	80.65	82.17	83.27	84.39	85.78	87.02	89.12	90.36	82.79
Nineteen	Girls.	82	77.02	78.05	80.04	81.00	81.73	83.20	84.03	85.06	87.12	88.95	89.98	82.30
Twenty	Girls.	65	78.75	80.12	82.00	83.25	84.83	85.79	86.83	87.50	89.00	89.81	91.75	84.99
Twenty-one	Girls.	32	80.60	81.40	82.70	84.20	85.26	86.33	87.40	89.20	90.40	91.40	92.40	86.97

TABLE No. 22.

THE GIRTH OF CHEST AT FULL EXPIRATION.

Age at Nearest Birthday.	Sex.	Number of Observations.	Value in Centimetres at the following Percentile Grades.										Average. Centimeters
			5	10	20	30	40	50	60	70	80	90	95	
Six	Boys.	679	52.00	54.06	55.12	56.09	56.92	57.60	58.35	59.24	60.12	61.85	64.20	57.26
	Girls.	743	52.36	53.46	54.97	55.90	56.91	57.48	58.28	59.15	60.28	61.84	63.64	57.00
Seven	Boys.	1715	54.09	55.10	56.36	57.28	58.15	58.92	59.77	60.65	61.83	63.74	65.55	58.68
	Girls.	1659	52.00	53.96	55.41	56.48	57.40	58.31	59.11	60.06	61.27	62.97	64.64	57.80
Eight	Boys.	2036	54.92	56.27	57.86	58.85	59.74	60.64	61.54	62.45	63.54	65.09	66.43	60.08
	Girls.	2036	54.06	55.05	56.53	57.62	58.53	59.49	60.47	61.44	62.63	64.28	65.40	59.01
Nine	Boys.	2121	56.66	57.79	59.24	60.37	61.29	62.15	63.06	64.21	65.44	67.18	68.91	61.78
	Girls.	1967	55.35	56.62	58.24	59.32	60.26	61.20	62.15	63.26	64.43	66.25	67.82	60.72
Ten	Boys.	1993	57.90	59.13	60.57	61.59	62.54	63.54	64.57	65.70	67.16	69.49	71.25	63.82
	Girls.	1894	56.33	57.71	59.34	60.55	61.61	62.47	63.48	64.61	65.92	67.79	69.92	62.07
Eleven	Boys.	1732	58.92	60.37	61.87	63.82	64.19	65.20	66.10	67.24	68.73	70.79	72.00	64.80
	Girls.	1664	57.66	58.92	60.52	61.93	63.12	64.28	65.49	66.65	67.75	69.98	71.90	63.80
Twelve	Boys.	1561	60.14	61.45	63.15	64.39	65.44	66.50	67.60	68.72	70.22	71.94	73.57	66.10
	Girls.	1619	59.36	61.02	62.88	64.01	65.21	66.38	67.75	69.27	70.98	73.16	75.19	66.26
Thirteen	Boys.	1215	61.52	62.78	64.46	65.82	67.14	68.33	69.48	70.58	72.00	74.51	76.28	67.92
	Girls.	1311	61.11	63.12	65.38	66.87	68.26	69.64	70.86	72.26	73.95	76.51	78.77	69.24
Fourteen	Boys.	929	63.24	64.55	66.64	67.98	69.16	70.32	71.76	73.47	75.22	77.59	79.80	70.33
	Girls.	1018	64.22	65.60	67.90	69.78	70.91	72.36	73.77	75.20	76.85	79.43	82.00	71.88
Fifteen	Boys.	499	65.33	67.15	69.35	70.84	72.34	73.89	74.95	76.44	78.36	80.95	83.65	73.37
	Girls.	657	66.20	68.39	70.67	72.13	73.26	74.78	76.30	77.94	80.10	82.38	84.65	74.61
Sixteen	Boys.	204	67.04	68.90	71.31	73.38	75.00	76.21	77.67	79.25	80.92	84.23	87.12	75.86
	Girls.	395	69.41	70.68	72.61	74.34	75.59	76.78	78.18	79.52	81.64	84.77	87.05	76.62
Seventeen	Boys.	80	68.00	72.33	74.33	76.00	77.44	78.33	79.33	81.50	82.67	84.00	86.00	77.61
	Girls.	201	70.68	72.51	74.16	75.74	76.98	78.27	79.76	81.12	83.00	85.36	88.00	78.01
Eighteen	Boys.	31	74.10	75.55	77.20	79.26	79.88	80.83	82.20	83.18	83.95	86.90	87.48	80.39
	Girls.	160	71.17	73.00	74.64	76.10	77.47	78.57	79.62	81.08	82.71	84.88	86.67	78.12
Nineteen	Girls.	81	68.50	70.35	71.84	73.55	74.60	76.42	77.77	78.96	80.23	82.45	85.48	75.91
Twenty	Girls.	67	70.07	70.74	73.13	74.41	75.20	76.50	77.64	79.32	80.15	82.38	82.94	76.09
Twenty-one	Girls.	32	71.17	70.20	73.20	74.40	75.40	76.66	77.73	79.70	80.65	84.40	86.20	76.06

318   *Trans. Acad. Sci. of St. Louis.*

TABLE No. 23.

The Girth of Chest Midway Between Full Inspiration and Full Expiration.*

Age at nearest Birthday.	Sex.	Number of Observations.	\multicolumn{10}{c}{Value of Centimetres at the following Percentile Grades.}	Average.
			5	10	20	30	40	50	60	70	80	90	95	Centimet's
Six	Boys.	677	54.02	55.59	56.92	57.98	58.64	59.32	60.04	60.91	61.93	63.72	65.62	59.65
	Girls.	741	53.56	54.93	56.34	57.27	58.21	58.87	59.60	60.42	61.54	63.13	64.84	58.34
Seven	Boys.	1208	55.91	57.04	58.31	59.26	60.31	60.89	61.73	62.67	63.82	65.59	67.46	60.62
	Girls.	1631	53.99	55.61	57.08	58.22	59.07	59.87	60.75	61.70	62.91	64.55	66.16	59.47
Eight	Boys.	2095	57.12	58.42	60.00	60.97	61.87	62.55	63.65	64.62	65.63	67.15	68.33	62.18
	Girls.	2040	55.93	56.98	58.40	59.47	60.40	61.31	62.32	63.23	64.36	65.98	67.19	60.81
Nine	Boys.	2120	58.99	59.98	61.42	62.52	63.46	64.33	65.23	66.31	67.51	69.17	70.80	63.90
	Girls.	1966	57.61	58.47	60.09	61.21	62.16	63.03	63.94	65.03	66.16	67.88	69.41	62.51
Ten	Boys.	1997	60.11	61.40	62.88	64.09	64.97	65.92	66.90	67.96	69.31	71.45	73.27	65.59
	Girls.	1893	57.84	58.98	60.59	61.77	62.72	63.44	65.44	65.53	67.79	69.47	71.36	63.02
Eleven	Boys.	1732	61.47	62.84	64.50	66.07	66.73	67.68	68.51	69.60	71.04	73.11	74.62	67.24
	Girls.	1654	60.00	61.09	62.71	64.05	65.25	66.36	67.62	68.56	69.68	71.91	73.85	65.85
Twelve	Boys.	1565	62.79	64.15	65.90	67.16	68.17	69.17	70.28	71.35	72.77	74.62	76.26	68.76
	Girls.	1624	61.50	63.16	65.08	66.25	67.38	68.49	69.76	71.19	72.89	75.15	77.21	68.34
Thirteen	Boys.	1228	63.95	65.49	67.59	68.58	69.86	71.01	72.16	73.37	74.88	77.23	79.14	70.61
	Girls.	1313	63.25	65.32	67.61	69.03	70.31	71.64	72.89	74.30	75.95	78.56	80.80	71.29
Fourteen	Boys.	925	66.22	67.57	69.53	70.94	72.20	73.38	74.85	76.44	78.09	80.56	82.77	73.27
	Girls.	1020	66.33	67.96	70.18	71.88	73.19	74.59	76.05	77.48	79.09	81.52	83.84	74.13
Fifteen	Boys.	498	68.44	70.29	72.51	73.98	75.48	76.94	78.26	79.85	81.29	84.23	86.60	75.56
	Girls.	659	68.69	70.73	72.92	74.41	75.76	76.92	78.47	80.32	81.97	84.12	86.45	76.78
Sixteen	Boys.	205	69.57	71.72	74.66	76.56	78.45	79.90	81.32	82.95	84.47	87.64	90.14	79.22
	Girls.	397	71.93	73.26	75.15	76.81	78.08	79.22	80.44	82.04	83.66	86.48	88.38	78.85
Seventeen	Boys.	90	70.87	75.61	78.08	79.72	81.10	82.20	83.28	85.05	86.48	88.64	90.17	81.39
	Girls.	206	73.30	74.82	76.81	78.24	79.45	80.72	82.00	83.23	85.03	87.42	89.64	80.39
Eighteen	Boys.	31	76.10	79.80	81.70	83.68	84.49	85.35	86.36	87.44	88.32	90.42	91.46	84.52
	Girls.	162	73.60	75.42	76.99	78.37	79.82	80.92	82.00	84.93	84.86	87.00	88.51	80.45
Nineteen	Girls.	82	72.76	74.20	75.94	77.27	78.16	79.81	80.90	82.01	83.67	85.70	87.73	79.10
Twenty	Girls.	66	74.41	75.43	77.56	78.83	80.01	81.04	82.23	83.41	84.57	86.09	87.34	80.54
Twenty-one	Girls.	32		75.80	77.95	79.30	80.33	81.49	82.56	84.45	85.62	87.90	89.30	81.51

* Obtained by adding the Girth of Chest at Full Inspiration to the Girth of Chest at Full Expiration and dividing the sum by 2.

TABLE No. 24.
THE LENGTH OF HEAD.

Age at Nearest Birthday.	Sex.	Number of Observations.	Value in Millimetres at the following Percentile Grades.											Millmett's Average.
			5	10	20	30	40	50	60	70	80	90	95	
Six.........	Boys.	606	169.19	170.88	173.50	175.75	177.34	178.86	180.39	181.50	184.08	186.95	189.23	178.89
	Girls.	606	163.69	166.42	169.44	170.94	172.89	174.34	175.80	177.40	178.87	180.98	183.93	173.46
Seven	Boys.	1493	168.59	170.99	173.87	175.80	177.85	179.42	181.02	182.71	184.86	187.34	190.03	178.54
	Girls.	1511	165.13	167.37	170.48	172.34	174.08	175.58	177.14	178.79	180.73	183.25	185.44	174.09
Eight.......	Boys.	2079	170.03	171.89	175.14	176.91	178.80	180.44	182.03	183.92	185.81	188.73	190.86	179.62
	Girls.	2125	165.55	168.42	171.07	173.01	175.25	176.41	178.17	180.15	181.48	184.18	185.90	175.18
Nine........	Boys.	1986	170.01	172.39	175.62	177.79	179.84	181.00	182.77	184.80	186.59	190.04	191.67	180.72
	Girls.	1884	166.58	169.93	172.61	174.77	176.08	177.80	179.54	180.76	182.59	185.55	187.53	176.39
Ten.........	Boys.	1912	171.50	174.37	177.09	179.10	180.20	181.51	183.22	184.99	186.70	189.85	191.82	181.45
	Girls.	1790	167.72	170.45	173.03	175.26	176.77	178.52	180.10	181.26	183.67	185.99	188.11	177.24
Eleven	Boys.	1654	172.66	175.00	177.70	179.99	181.38	182.86	184.62	186.17	188.24	190.93	193.38	182.37
	Girls.	1560	167.47	170.73	173.69	175.73	177.82	179.76	181.00	182.71	184.84	187.45	190.35	178.08
Twelve	Boys.	1576	172.79	175.22	178.05	180.32	181.76	183.08	184.78	186.41	188.63	191.85	194.65	182.84
	Girls.	1516	169.79	172.32	175.42	177.44	179.07	180.57	182.22	183.88	185.84	189.00	191.11	179.50
Thirteen....	Boys.	1207	173.60	176.07	179.00	180.85	182.54	184.21	185.91	187.63	189.92	192.53	195.10	183.84
	Girls.	1187	171.17	173.74	177.31	179.38	180.86	182.48	184.33	185.97	188.12	190.69	192.82	181.44
Fourteen ...	Boys.	890	175.19	177.88	181.06	183.10	184.73	186.24	187.95	189.60	191.89	195.00	196.79	186.93
	Girls.	1008	173.03	175.93	179.05	181.09	182.87	184.71	186.02	187.65	189.90	192.68	195.08	183.41
Fifteen.....	Boys.	502	176.67	178.98	182.18	183.84	185.73	187.69	190.00	191.00	192.94	196.26	199.09	187.01
	Girls.	649	174.49	177.46	180.59	182.47	184.33	185.67	187.67	189.46	191.34	194.32	195.98	185.12
Sixteen.....	Boys.	191	178.83	180.48	183.83	185.74	188.21	190.04	191.56	194.01	195.53	198.48	200.73	189.06
	Girls.	400	175.80	179.00	181.82	184.21	186.00	187.70	189.50	191.40	193.64	196.08	198.00	186.84
Seventeen ...	Girls.	221	179.02	181.03	183.62	185.63	187.18	188.60	190.03	191.34	193.85	197.38	199.59	188.14
Eighteen	Girls.	161	179.36	181.22	184.12	185.46	187.11	188.41	190.04	190.98	192.73	195.99	198.98	187.97
Nineteen....	Girls.	77	180.70	181.85	183.80	184.92	185.97	188.31	189.30	190.70	192.80	195.66	199.06	187.91

TABLE No. 25.
THE WIDTH OF HEAD.

Age at Nearest Birthday.	Sex.	Number of Observations.	Value in Millimetres at the following Percentile Grades.										Millmeir's Average.
			5	10	20	30	40	50	60	70	80	90	95	
Six	Boys. Girls.	573 609	134.70 131.74	138.01 133.93	140.08 136.25	141.45 138.17	142.73 139.91	143.76 140.83	145.11 142.31	146.20 143.64	148.01 145.44	150.60 147.93	152.72 150.25	152.72 143.29 140.27
Seven	Boys. Girls.	1571 1505	136.31 133.66	138.87 135.56	140.80 137.64	142.33 139.61	143.46 140.80	144.98 142.10	145.93 143.25	147.25 144.74	149.26 146.13	151.51 148.78	153.54 150.70	144.37 141.40
Eight	Boys. Girls.	1997 1985	136.78 133.70	139.38 136.03	141.60 138.61	143.14 140.39	144.48 141.48	145.63 142.81	146.70 144.24	148.43 145.58	150.42 147.01	152.84 150.13	155.08 151.73	145.30 142.31
Nine	Boys. Girls.	1962 1914	137.44 134.82	140.04 136.80	141.97 139.88	143.62 141.05	145.07 142.40	146.08 143.58	147.44 144.96	149.20 146.61	150.89 148.01	153.56 150.61	155.49 152.63	145.87 143.04
Ten	Boys. Girls.	1827 1803	138.30 134.94	140.53 137.49	142.57 140.31	144.22 141.66	145.72 143.15	147.11 144.54	148.58 145.64	150.27 146.92	151.76 149.11	154.15 151.49	156.02 154.09	146.59 143.75
Eleven	Boys. Girls.	1561 1541	139.39 136.71	141.29 139.27	143.40 141.30	145.19 142.82	146.31 144.30	147.50 145.54	148.84 146.74	150.57 148.40	152.27 150.25	154.94 152.78	157.26 154.94	147.29 145.05
Twelve	Boys. Girls.	1531 1460	140.32 136.83	142.19 139.22	144.57 142.19	145.80 143.67	146.91 145.16	148.36 146.31	150.07 147.70	151.25 149.24	152.91 150.88	155.40 153.38	156.99 155.63	147.98 145.64
Thirteen	Boys. Girls.	1175 1167	140.41 138.49	142.32 140.40	144.88 142.61	146.03 144.26	147.48 145.76	149.01 147.15	150.68 148.53	152.10 150.22	153.65 152.12	156.25 154.82	158.79 156.79	148.73 146.78
Fourteen	Boys. Girls.	873 927	141.32 139.57	142.78 141.47	145.19 143.68	146.88 145.46	148.37 146.83	150.04 148.25	151.03 149.71	152.40 151.14	154.44 153.13	157.47 155.88	160.26 157.98	149.50 147.90
Fifteen	Boys. Girls.	469 658	140.83 140.17	143.06 141.99	145.46 144.29	147.02 146.03	148.68 147.32	150.41 148.71	151.83 150.09	153.27 151.57	154.80 153.69	156.96 156.36	159.79 158.01	149.63 148.29
Sixteen	Boys. Girls.	195 396	144.22 140.39	145.32 142.28	147.00 145.00	148.50 146.69	150.25 148.39	151.35 149.67	152.65 150.89	154.03 152.48	155.88 154.39	157.72 156.53	160.71 159.05	150.98 148.95
Seventeen	Boys. Girls.	75 221	142.75 142.12	146.50 143.39	148.43 146.25	151.25 147.72	151.00 149.03	152.50 150.43	153.57 151.71	155.50 153.44	157.60 155.31	159.75 156.81	165.25 160.24	152.09 150.04
Eighteen	Girls.	165	141.42	142.81	145.27	146.65	148.46	150.16	151.23	152.50	153.91	156.62	158.69	149.09
Nineteen	Girls.	79	140.95	141.98	145.26	146.90	148.33	150.17	151.13	153.43	154.77	156.70	159.06	149.11
Twenty	Girls.	73	139.33	142.26	144.20	145.18	146.46	147.70	149.80	151.62	152.90	155.99	159.70	147.82

TABLE No. 26.

THE HEIGHT OF FACE FROM ROOT OF NOSE TO POINT OF CHIN.

Age at Nearest Birthday.	Sex.	Number of Observations.	Value in Millimetres at the following Percentile Grades.										Average. Millimetres
			5	10	20	30	40	50	60	70	80	90	95	
Six............	Boys.	604	82.78	85.54	88.32	90.15	91.50	93.20	95.29	97.94	100.96	104.83	107.53	93.72
	Girls.	612	82.14	83.62	85.98	87.87	89.15	90.89	93.14	95.22	97.69	100.96	105.05	91.42
Seven.........	Boys.	1580	85.73	88.23	90.44	92.12	93.96	95.57	97.39	99.86	102.36	105.83	109.78	95.87
	Girls.	1509	83.26	85.68	88.28	90.41	91.76	93.20	95.17	96.97	99.78	103.49	106.63	93.77
Eight..........	Boys.	2057	87.31	90.21	92.63	94.82	96.07	97.78	99.81	101.28	104.14	107.72	110.65	97.98
	Girls.	2011	86.22	88.34	90.78	92.66	94.24	95.60	97.15	98.97	101.61	105.49	108.77	95.65
Nine...........	Boys.	2011	90.06	91.43	94.14	95.93	97.60	100.77	100.77	103.03	105.46	109.59	112.35	99.51
	Girls.	1898	88.60	90.42	92.63	94.59	95.97	97.60	99.43	101.04	103.68	107.42	111.22	97.85
Ten............	Boys.	1868	91.19	93.32	95.74	97.58	99.14	100.73	102.95	104.87	106.99	110.72	113.16	101.06
	Girls.	1820	90.04	91.72	94.18	95.89	97.53	99.02	100.94	103.00	105.44	109.72	112.17	99.39
Eleven.........	Boys.	1660	92.52	94.94	97.52	99.54	100.96	103.23	105.16	107.35	110.35	113.90	117.05	103.37
	Girls.	1572	91.94	93.79	96.16	98.00	99.75	101.20	102.87	104.97	107.67	111.10	114.60	101.44
Twelve.........	Boys.	1568	94.74	96.59	98.96	100.80	102.43	104.45	105.97	108.71	110.83	115.47	120.03	104.25
	Girls.	1523	91.76	94.52	97.68	99.94	101.92	103.53	105.23	107.30	110.36	113.65	117.38	103.46
Thirteen.......	Boys.	1205	95.58	97.98	100.61	102.66	104.86	106.34	108.71	110.61	112.91	116.66	120.70	106.24
	Girls.	1231	95.70	98.15	100.66	102.54	104.21	105.71	107.85	110.19	112.21	115.60	118.59	105.92
Fourteen.......	Boys.	893	96.55	99.70	102.51	104.77	106.17	108.51	110.40	112.35	115.23	119.43	123.04	108.43
	Girls.	897	96.99	99.94	102.38	104.36	105.76	107.72	109.61	111.45	114.27	118 85	121.86	107.87
Fifteen.........	Boys.	479	100.38	102.69	105.00	106.98	109.97	111.52	113.18	115.73	118.02	122.55	126.36	111.63
	Girls.	672	98.24	100.68	103.31	105.80	108.07	109.40	110.96	112.75	115.34	120.24	122.74	109.79
Sixteen........	Boys.	191	103.31	105.02	107.89	110.48	112.49	113.73	115.62	117.43	120.52	123.73	126.96	113.63
	Girls.	403	100.42	102.02	105.07	106.77	108.21	109.83	112.17	114.00	116.87	121.00	125.43	110.48
Seventeen.....	Boys.	78	102.45	105.80	111.04	113.30	116.80	118.50	120.47	121.58	124.20	130.40	135.10	117.56
	Girls.	223	98.79	101.66	104.40	105.82	108.24	110.50	112.59	114.18	116.74	120.18	123.85	110.04
Eighteen......	Girls.	163	100.23	101.72	104.94	106.36	108.42	110.04	112.32	112.84	115.27	119.34	122.85	109.77
Nineteen......	Girls.	80	101.00	102.17	103.76	105.60	106.75	108.34	110.20	111.80	114.00	117.00	121.00	108.72
Twenty........	Girls.	73	102.16	103.08	105.53	107.15	109.17	110.30	111.63	112.85	115.40	120.18	121.35	110.26

322   Trans. Acad. Sci. of St. Louis.

TABLE No. 27.
THE WIDTH OF FACE.

Value in Millimetres at the following Percentile Grades.

Age at Nearest Birthday.	Sex.	Number of Observations.	5	10	20	30	40	50	60	70	80	90	95	Average. Millimer's
Six	Boys.	608	105.64	110.25	112.85	115.23	116.77	118.20	119.64	120.74	122.50	125.44	127.78	117.24
	Girls.	608	105.28	109.20	111.19	112.64	114.51	115.89	116.88	118.40	120.38	123.10	126.51	115.21
Seven	Boys.	1637	108.08	110.83	113.43	115.38	116.68	118.40	119.98	121.04	122.89	125.59	128.38	117.78
	Girls.	1516	107.65	110.25	112.15	113.75	115.21	116.38	117.80	119.57	120.87	123.36	126.09	116.11
Eight	Boys.	2062	110.02	111.98	114.87	116.43	118.29	119.75	120.69	122.00	123.91	126.27	128.72	118.82
	Girls.	2065	108.16	110.57	112.57	114.66	116.01	117.45	118.84	120.32	121.66	124.60	126.95	117.63
Nine	Boys.	1927	110.05	112.10	115.49	117.59	119.70	120.75	122.14	123.67	125.51	128.08	130.62	119.91
	Girls.	1905	108.20	110.62	113.55	115.63	117.17	118.81	120.24	121.36	122.99	125.59	127.83	118.02
Ten	Boys.	1922	110.64	113.35	116.90	119.14	120.49	121.63	122.98	124.48	126.16	128.73	131.00	121.60
	Girls.	1829	108.77	112.22	115.21	117.16	118.91	120.42	121.55	123.09	124.98	127.51	129.40	119.49
Eleven	Boys.	1666	112.43	115.26	118.44	120.44	121.81	123.22	124.85	125.95	127.82	130.59	133.04	122.73
	Girls.	1604	111.45	113.70	116.66	118.98	120.59	121.71	122.84	124.29	126.87	129.15	132.07	121.24
Twelve	Boys.	1525	112.52	115.85	120.05	121.75	123.21	124.59	125.85	127.44	129.45	131.66	134.61	123.81
	Girls.	1526	111.89	114.85	118.45	120.55	121.84	123.07	124.46	126.13	127.88	130.89	132.77	122.44
Thirteen	Boys.	1213	114.96	118.24	120.96	122.95	124.63	126.02	127.38	128.92	130.65	133.19	136.40	125.83
	Girls.	1248	113.10	116.40	119.71	121.53	123.38	124.86	126.28	127.87	129.76	132.35	134.78	124.15
Fourteen	Boys.	898	116.53	120.13	122.04	124.30	125.78	127.82	128.93	130.63	132.65	135.39	137.51	126.81
	Girls.	997	114.09	118.39	121.25	123.37	125.08	126.51	128.08	129.76	131.35	133.80	135.82	125.67
Fifteen	Boys.	485	118.18	120.40	122.88	125.16	127.13	129.03	130.70	132.40	135.03	137.45	140.34	128.32
	Girls.	678	118.57	120.75	123.68	125.39	126.86	128.30	129.66	131.08	132.96	135.88	138.41	127.85
Sixteen	Boys.	193	120.88	123.29	125.40	127.38	129.22	130.82	132.70	134.01	135.77	138.78	140.91	130.27
	Girls.	409	120.08	122.98	125.38	127.03	128.50	130.18	131.45	132.72	134.47	136.90	139.28	129.47
Seventeen	Girls.	219	122.19	123.99	126.76	128.05	130.04	131.39	132.55	134.22	135.78	138.28	140.03	130.63
Eighteen	Girls.	163	124.03	125.41	128.07	130.21	130.92	131.97	133.24	134.34	135.84	138.30	139.93	131.42

TABLE No. 28.
The Height of Face from Hair-Line to Point of Chin.

Value in Millimetres at the following Percentile Grades.

Age at Nearest Birthday.	Sex.	Number of Observations.	5	10	20	30	40	50	60	70	80	90	95	Average. Millimetrs
Six	Boys.	611	141.32	144.06	147.34	149.93	150.90	153.02	155.10	157.23	159.94	162.68	165.73	152.68
	Girls.	609	137.55	140.46	143.49	146.87	147.74	150.45	152.61	155.01	157.40	161.34	164.51	150.16
Seven	Boys.	1621	141.62	145.07	148.87	151.39	153.55	155.33	156.97	159.41	161.38	165.36	168.35	154.57
	Girls.	1486	141.16	144.31	147.24	150.21	152.08	154.19	155.90	158.16	160.47	163.36	165.84	153.36
Eight	Boys.	2012	144.88	148.35	151.75	154.32	156.10	158.04	160.20	161.72	164.48	168.36	170.86	157.50
	Girls.	1965	143.38	146.14	150.03	152.07	153.89	155.60	157.45	160.08	162.13	165.61	168.76	155.19
Nine	Boys.	1997	146.61	150.18	153.14	155.75	158.04	160.25	162.06	164.40	166.84	170.46	173.14	159.43
	Girls.	1989	145.83	148.77	150.90	153.72	155.73	157.76	160.18	162.06	164.73	168.23	170.85	157.44
Ten	Boys.	1909	148.67	151.85	155.35	157.59	160.27	161.70	163.71	165.89	168.86	172.25	175.25	161.37
	Girls.	1835	147.07	150.37	154.28	156.43	158.49	160.56	162.57	164.90	167.40	170.85	174.37	160.04
Eleven	Boys.	1653	150.47	153.75	157.29	160.04	162.05	164.11	165.72	168.10	170.80	174.94	177.72	163.39
	Girls.	1581	149.29	152.43	156.32	158.44	161.09	163.40	165.35	168.89	170.17	173.13	176.08	162.55
Twelve	Boys.	1577	152.33	155.49	159.36	161.38	163.51	165.62	167.55	170.41	172.67	176.39	180.16	165.23
	Girls.	1510	151.83	155.14	158.77	161.77	163.46	165.95	168.41	170.34	172.46	176.61	180.06	165.29
Thirteen	Boys.	1211	154.68	157.30	161.24	163.94	165.94	168.16	170.31	172.33	175.40	178.74	181.49	167.62
	Girls.	1220	154.62	158.00	162.00	164.35	166.38	169.15	171.32	173.02	176.08	179.96	182.64	168.35
Fourteen	Boys.	896	155.98	158.90	163.66	166.35	168.65	171.04	173.56	175.98	178.86	182.59	185.95	170.49
	Girls.	998	156.99	160.43	164.85	167.19	169.73	172.00	174.68	176.91	179.64	183.01	185.77	171.38
Fifteen	Boys.	482	160.37	163.24	167.14	170.33	172.68	175.08	177.46	179.85	182.60	186.92	190.17	174.30
	Girls.	656	161.98	164.56	167.28	170.55	172.78	174.85	176.63	178.86	181.34	185.03	187.72	174.20
Sixteen	Boys.	193	165.22	168.38	171.66	173.86	175.75	177.93	180.36	183.01	185.42	191.23	194.68	178.19
	Girls.	395	163.75	166.35	170.15	172.48	174.67	176.81	179.05	181.38	184.80	188.12	190.12	176.28
Seventeen	Girls.	201	162.03	167.41	172.31	174.90	177.17	179.41	180.07	183.59	186.20	190.38	194.48	
Eighteen	Girls.	139	167.48	170.22	174.95	177.68	180.33	181.93	183.57	185.55	188.07	192.03	195.02	
Nineteen	Girls.	69	166.50	168.95	172.90	175.68	177.72	180.08	181.35	184.32	187.07	190.55	194.28	
Twenty	Girls.	72	169.30	171.60	174.08	175.52	177.16	179.00	181.07	182.88	185.87	188.90	192.20	

CHAPTER VI.

SEXUAL DIFFERENCES IN GROWTH.

When the curves of growth in weight, height standing, height sitting, span of arms and girth of chest are drawn on the same system of co-ordinates, as has been done in Plates XXV to XXIX inclusive, the attention of the observer is arrested by the extraordinary difference in the development of girls and boys during the period of prepubertal acceleration. Girls enter this time of rapid growth at age 11 or 12, two years earlier than boys, and during several years are larger than boys of the same age. The period during which girls are larger than boys does not correspond exactly with the period of accelerated development, but begins and ends a little later. The ages at which girls begin and cease to be larger than boys are given in the following table. The ages

TABLE No. 29.

AGES AT WHICH GIRLS BEGIN AND CEASE TO BE LARGER THAN BOYS.

DIMENSION.	Age at which Girls begin to be larger than Boys. Percentile Grades.			Age at which Girls cease to be larger than Boys. Percentile Grades.		
	25	50	75	25	50	75
Weight............	$12\tfrac{6}{12}$ yrs.	$12\tfrac{2}{12}$ yrs.	$11\tfrac{6}{12}$ yrs.	$16\tfrac{3}{12}$ yrs.	$15\tfrac{6}{12}$ yrs.	$14\tfrac{8}{12}$ yrs.
Height standing.....	$11\tfrac{8}{12}$ "	$11\tfrac{7}{12}$ "	$11\tfrac{3}{12}$ "	$15\tfrac{8}{12}$ "	$15\tfrac{3}{12}$ "	$14\tfrac{0}{12}$ "
Height sitting........	$11\tfrac{11}{12}$ "	$11\tfrac{6}{12}$ "	$11\tfrac{5}{12}$ "	$16\tfrac{11}{12}$ "	$16\tfrac{9}{12}$ "	$16\tfrac{2}{12}$ "
Span of arms.........	$12\tfrac{7}{12}$ "	$11\tfrac{11}{12}$ "	$11\tfrac{8}{12}$ "	$15\tfrac{3}{12}$ "	$14\tfrac{5}{12}$ "	$14\tfrac{2}{12}$ "
Girth of chest........	$12\tfrac{9}{12}$ "	$12\tfrac{4}{12}$ "	12 "	$16\tfrac{3}{12}$ "	15 "	$15\tfrac{5}{12}$ "
Height of face from hair line to point of chin...............		$11\tfrac{9}{12}$ yrs.			15 "	

are nearly the same for the same percentile grade in all five dimensions. An examination of plates XXX to XXXVI shows that the sexual difference just noted is not present in expansion of chest, or in strength of squeeze, or in any head or face measurement except height of face from hair-line to point of chin. Boys have therefore a larger expansion of the chest, greater strength of squeeze and greater length and width of head and height and width of face than girls throughout their period of growth.

An interesting comparison can be made of the duration of the period in which girls are larger than boys. It appears

DURATION OF THE PERIOD DURING WHICH GIRLS ARE LARGER THAN BOYS.

DIMENSION.	Percentile Grades.		
	25	50	75
Weight..........................	$3\frac{2}{12}$ yrs.	$3\frac{4}{12}$ yrs.	$3\frac{2}{12}$ yrs.
Height standing.................	4 "	$3\frac{10}{12}$ "	$3\frac{3}{12}$ "
Height sitting..................	5 "	5 "	$4\frac{9}{12}$ "
Span of arms....................	$3\frac{1}{12}$ "	$2\frac{8}{12}$ "	$2\frac{6}{12}$ "
Girth of chest..................	$3\frac{6}{12}$ "	$2\frac{4}{12}$ "	$3\frac{1}{12}$ "

that the period is shortest in span of arms and is considerably longer in height sitting than in any other dimension.

The age at which girls begin to be larger than boys differs at different percentile grades, big girls (75 percentile grades) beginning to be larger than big boys at an earlier age than that at which small girls begin to exceed small boys. A difference is seen also in the duration of the period in which girls are larger than boys: the small girls keep their superiority during a longer time than the larger girls.

Sexual differences are further displayed in Plates XXXVII to XLI, inclusive, in which the percentile curves of both sexes are drawn one a short distance under the other, and the points at which girls begin and cease to be heavier than boys joined by heavy unbroken lines. The early superiority of large girls and the relative early loss of their superiority is seen in all the plates. The fact that the period during which big girls are larger than big boys is shorter than the period during which little girls are larger than little boys is very clearly demonstrated by Plates XXXVIII, Height Standing, XXXIX, Height Sitting and XL, Span of Arms; it can be seen also in Plate XXXVII, Weight, and XLI, Girth of Chest, if the ninety and the five percentile grades, in which the small number of observations has probably led to error, are neglected.

CHAPTER VII.

THE RATE OF GROWTH.

The Absolute Annual Increase is the gain in weight or height, etc., during the twelve preceding months; thus, the absolute annual increase in height at age 7 is the gain in weight during the twelve months from age 6 to age 7, obtained by subtracting the average or median weight at age 6 from that at age 7.

The absolute annual increase of height standing, weight and span of arms is shown in Plate XLII and Tables No. 30, 31 and 32. In all three, the maximum for girls is at age 13 and the maximum for boys at age 15. The same may be said of the curves for height sitting, girth of chest and strength of squeeze, in Plate XLIII, from Tables No. 33, 34, 44 and 45, although the curves are less regular, owing to the observations being more difficult than those from which the preceding plate was constructed. In all six curves, the small number of observations at age 17 and 18 cause the median values at those ages to be less reliable than at other ages.

The Relative Annual Increase is the increase for any year divided by the average value at that year; thus, the relative annual increase in weight at age 7 is the difference between the average weight at age 6 and age 7 divided by the average weight at age 6. The relative annual increase gives a truer idea of growth than the absolute annual increase, because the latter value is entangled with the size of the individual measured. The absolute increase is commonly greater in a big boy than in a small boy, and yet the rate of growth may be the same. The relative annual increase is free of such errors.

The relative annual increase in strength of squeeze, weight, height standing, height sitting, span of arms and girth of chest is drawn in Plate XLIV. The gain in weight during the prepubertal acceleration is extraordinary in both girls and boys, as is the rapid fall immediately thereafter. The quickness of growth in height standing of boys is slightly greater

TABLE No. 30.

The Absolute Annual Increase in Height Standing.

Age at nearest Birthday.	Sex.	Value in Centimetres at the following Percentile Grades.										Increase in average Height. Centimeter's
		5	10	20	30	40	50	60	70	80	90	95	
Six to Seven	Boys.	3.90	4.06	4.98	5.08	5.07	5.25	5.41	5.27	5.39	5.43	5.07	5.08
	Girls.	4.73	4.83	4.96	5.00	5.11	5.34	5.63	5.71	5.84	5.38	5.04	5.28
Seven to Eight	Boys.	4.01	4.72	4.92	5.20	5.47	5.50	5.42	5.59	5.54	5.49	5.66	5.10
	Girls.	5.25	5.02	5.39	5.31	5.51	5.31	5.13	5.10	5.45	5.41	5.85	5.41
Eight to Nine	Boys.	6.21	5.85	5.14	4.89	4.94	5.09	5.02	5.15	5.20	5.17	5.01	5.22
	Girls.	4.99	5.26	5.27	5.39	5.36	5.36	5.37	5.12	5.35	6.18	5.63	5.31
Nine to Ten	Boys.	4.15	3.81	4.29	4.50	4.60	4.58	4.73	4.67	4.89	5.24	5.39	4.52
	Girls.	4.26	4.22	4.55	4.56	4.56	4.74	4.80	5.28	4.71	5.89	5.42	4.76
Ten to Eleven	Boys.	4.49	4.67	4.63	4.55	4.77	4.99	4.87	4.70	4.80	6.04	5.30	4.97
	Girls.	3.52	4.07	4.02	4.62	4.76	4.75	4.92	5.05	6.73	5.30	5.85	4.76
Eleven to Twelve	Boys.	4.12	4.00	4.15	4.28	4.12	4.13	4.56	4.70	4.72	4.84	5.08	4.37
	Girls.	5.11	5.25	5.37	5.39	5.47	5.94	6.13	6.42	6.52	7.00	7.50	5.92
Twelve to Thirteen	Boys.	2.71	3.55	4.13	4.28	4.52	4.72	4.71	5.07	5.49	6.52	6.08	4.70
	Girls.	6.10	6.43	6.31	6.44	6.68	6.65	6.88	7.39	7.32	6.87	6.44	7.42
Thirteen to Fourteen	Boys.	5.74	5.49	4.91	5.05	5.21	5.57	5.55	6.25	6.23	5.64	6.97	5.67
	Girls.	4.86	4.57	5.84	5.89	6.02	5.75	5.44	4.69	4.36	4.28	3.36	4.31
Fourteen to Fifteen	Boys.	3.84	4.06	5.20	5.70	6.15	6.39	6.80	6.93	7.76	8.81	7.56	6.32
	Girls.	6.09	6.54	5.14	4.67	4.03	3.88	3.19	3.23	3.29	2.46	2.84	4.20
Fifteen to Sixteen	Boys.	5.48	5.69	6.43	6.36	6.80	6.02	5.61	5.28	5.07	3.02	3.11	5.37
	Girls.	3.89	2.25	2.23	2.14	2.24	2.21	2.65	2.37	2.10	3.16	2.81	2.48
Sixteen to Seventeen	Boys.	8.48	7.52	5.79	5.23	4.75	4.73	5.05	4.73	3.88	3.02	3.50	4.86
	Girls.	1.43	2.01	1.68	1.75	1.25	1.37	1.00	1.05	1.18	0.30	0.82	1.81
Seventeen to Eighteen	Boys.	0.67	1.01	1.00	0.56	0.77	4.50	0.03	0.24	0.48	0.22	1.14	5.28
	Girls.						0.10						0.09

TABLE No. 31.

THE ABSOLUTE ANNUAL INCREASE IN WEIGHT.

Age at Nearest Birthday	Sex	Value in Kilogrammes at the following Percentile Grades.										Average Increase Kilogram.
		5	10	20	30	40	50	60	70	80	90	95	
Six to Seven	Boys.	1.66	1.48	1.46	1.67	1.68	1.81	1.92	2.09	2.09	2.45	2.72	2.11
	Girls.	1.16	1.35	1.49	1.61	1.79	1.82	1.90	2.10	2.11	2.30	2.76	1.89
Seven to Eight	Boys.	1.32	1.59	1.83	1.90	2.07	2.21	2.36	2.45	2.54	2.63	2.65	2.11
	Girls.	1.67	1.66	1.76	1.78	1.93	2.04	2.21	2.18	2.33	2.72	2.63	2.06
Eight to Nine	Boys.	2.07	2.21	2.15	2.19	2.20	2.35	2.33	2.42	2.43	2.26	3.04	2.28
	Girls.	1.62	1.79	2.01	2.09	2.08	2.19	2.22	2.46	2.56	2.52	2.86	2.20
Nine to Ten	Boys.	1.41	1.54	1.82	1.98	2.23	2.14	2.41	2.41	2.65	3.03	2.57	2.26
	Girls.	1.48	1.78	1.84	2.17	2.32	2.41	2.46	2.56	2.73	2.69	2.92	2.41
Ten to Eleven	Boys.	2.34	2.16	2.32	2.54	2.50	2.72	2.67	2.77	2.93	2.96	2.21	2.68
	Girls.	1.95	2.08	2.29	2.12	2.29	2.35	2.66	2.79	2.98	4.08	4.19	2.66
Eleven to Twelve	Boys.	2.16	2.22	2.16	2.14	2.40	2.33	2.38	2.63	2.58	3.07	3.46	2.51
	Girls.	2.35	2.34	2.62	2.92	3.02	3.45	3.56	3.86	4.11	4.61	6.43	3.51
Twelve to Thirteen	Boys.	1.82	1.90	2.09	2.37	2.39	2.84	2.98	3.31	3.88	5.19	5.95	3.10
	Girls.	2.96	3.18	3.65	3.97	4.37	4.75	5.31	5.82	6.10	6.51	5.47	4.83
Thirteen to Fourteen	Boys.	2.73	3.20	3.56	3.49	3.68	3.73	4.61	5.08	6.37	5.92	6.58	3.83
	Girls.	2.31	3.36	3.68	4.34	4.62	4.65	4.49	4.68	4.84	4.45	4.74	3.80
Fourteen to Fifteen	Boys.	3.74	3.38	3.32	3.94	4.67	5.51	5.78	6.19	6.55	6.79	6.33	5.78
	Girls.	4.98	4.40	5.42	4.36	4.39	4.20	3.75	3.17	3.06	2.92	3.06	4.40
Fifteen to Sixteen	Boys.	1.96	2.82	5.53	5.72	6.34	6.34	6.67	6.71	5.66	6.76	6.87	5.38
	Girls.	3.38	3.55	3.82	4.11	3.61	3.11	3.51	3.43	4.17	4.45	3.51	3.56
Sixteen to Seventeen	Boys.	7.22	7.58	6.79	5.77	4.79	3.48	3.25	3.36	3.18	0.04	1.66	4.07
	Girls.	4.41	3.57	2.15	2.08	2.43	2.56	2.72	2.53	1.48	0.84		2.36
Seventeen to Eighteen	Girls.	1.38	1.53	0.73	0.47	0.76	0.58	0.05	0.09	0.64	0.61	1.11	

TABLE No. 32.
THE ABSOLUTE ANNUAL INCREASE IN SPAN OF ARMS.

Age at Nearest Birthday.	Sex.	Value in Centimetres at the following Percentile Grades.										Average of Increase
		5	10	20	30	40	50	60	70	80	90	95	Centimetr's
Six to Seven	Boys.	4.47	4.85	5.17	5.68	5.31	5.51	5.49	4.94	5.59	5.26	5.18	5.47
	Girls.	4.71	4.85	4.73	4.32	4.77	6.15	5.93	6.21	5.52	5.49	5.69	5.40
Seven to Eight	Boys.	5.05	5.45	5.43	5.44	5.55	5.65	6.03	5.78	5.91	6.17	6.31	5.65
	Girls.	5.38	6.64	6.19	5.80	5.95	5.02	5.68	5.31	5.20	6.15	6.14	5.97
Eight to Nine	Boys.	5.48	5.43	5.10	4.96	5.01	4.94	4.93	5.33	5.10	5.41	5.96	5.11
	Girls.	5.08	4.00	5.40	5.50	5.28	6.15	5.37	5.43	6.35	5.16	5.85	5.30
Nine to Ten	Boys.	3.81	3.97	4.51	4.76	4.89	4.94	5.32	5.49	5.63	5.82	5.29	5.04
	Girls.	4.64	4.82	4.44	4.52	4.73	4.90	5.02	5.17	5.48	6.85	6.14	5.12
Ten to Eleven	Boys.	4.89	4.62	4.71	4.71	4.76	4.98	5.11	5.16	5.01	5.01	5.51	4.91
	Girls.	4.54	4.60	4.59	4.99	5.31	5.38	6.40	5.63	6.84	6.11	6.11	5.49
Eleven to Twelve	Boys.	4.09	4.23	4.59	4.71	4.90	4.89	4.72	4.71	5.00	5.51	5.94	5.47
	Girls.	4.67	5.07	5.68	5.71	5.98	6.04	6.33	6.57	6.58	6.40	7.36	5.83
Twelve to Thirteen	Boys.	3.78	4.35	4.24	4.82	4.59	4.85	5.13	5.70	6.19	6.78	6.43	4.49
	Girls.	7.50	7.15	6.79	7.17	7.29	7.23	7.75	7.44	7.23	7.08	6.19	7.12
Thirteen to Fourteen	Boys.	5.88	5.71	5.97	5.88	6.20	6.29	6.68	5.95	6.03	6.24	7.70	6.19
	Girls.	4.65	5.43	6.12	6.29	5.96	5.73	5.11	5.20	5.11	3.98	4.15	5.39
Fourteen to Fifteen	Boys.	2.75	4.30	5.60	6.02	6.73	7.17	7.13	8.00	9.35	9.97	10.21	7.15
	Girls.	6.08	5.56	4.48	4.15	3.48	2.65	3.16	3.05	2.69	3.17	3.54	3.80
Fifteen to Sixteen	Boys.	4.86	4.86	5.98	6.48	6.53	6.52	6.67	5.88	5.03	3.72	3.02	5.53
	Girls.	2.45	2.07	3.06	2.49	2.29	3.10	2.28	1.83	1.87	1.84	1.91	2.13
Sixteen to Seventeen	Boys.	8.68	9.28	7.36	6.09	5.01	4.27	2.92	4.22	3.56	3.36	2.95	4.60
	Girls.	1.07	0.85	0.08		0.17	0.31	0.08	0.27	0.11			0.50
Seventeen to Eighteen	Boys.	11.70	8.60	6.74	6.00	5.18	5.92	6.57	4.37	4.74	4.55	2.90	6.75
	Girls.	1.10	1.28	1.11	1.47	2.17	1.76	1.77	1.99	1.93	1.97	1.63	1.46

TABLE No. 33.

THE ABSOLUTE ANNUAL INCREASE IN HEIGHT SITTING.

Age at Nearest Birthday.	Sex.	Value In Centimetres at the following Percentile Grades.									Increase in Average Height Sitting.
		5	10	20	30	40	50	60	70	80	90	95	Centimetr's
Six to Seven.......	Boys.	2.13	2.00	2.01	2.10	2.25	2.45	2.46	2.35	2.05	2.07	0.61	2.01
	Girls.	2.22	2.37	2.48	2.21	2.39	2.33	2.48	2.63	2.50	2.40	2.45	2.45
Seven to Eight......	Boys.	1.80	1.88	1.98	2.14	2.10	1.99	1.89	1.92	1.99	1.63	1.13	1.42
	Girls.	2.81	2.35	2.11	2.15	2.21	2.52	1.94	1.70	1.92	2.13	2.07	2.17
Eight to Nine.......	Boys.	1.89	1.78	1.89	1.76	1.77	1.87	2.06	2.13	2.14	2.16	2.11	1.99
	Girls.	1.57	1.99	1.99	2.16	1.99	1.23	2.30	2.51	2.36	2.28	2.54	2.19
Nine to Ten.........	Boys.	1.60	2.33	1.76	1.95	2.08	2.10	2.05	2.02	2.07	2.29	0.36	2.52
	Girls.	1.67	1.87	1.93	1.92	2.06	2.69	2.09	2.15	2.21	2.20	2.21	2.03
Ten to Eleven	Boys.	1.53	2.13	1.94	1.82	1.86	1.91	1.86	1.96	1.89	1.77	3.84	1.42
	Girls.	1.69	1.41	1.04	0.86	1.83	1.74	1.71	1.88	2.02	2.29	2.35	1.84
Eleven to Twelve...	Boys.	2.39	1.73	1.29	1.62	1.61	1.59	1.72	1.80	1.81	1.77	2.18	1.88
	Girls.	0.91	2.19	2.98	3.08	2.50	2.60	2.79	2.90	2.78	3.16	2.72	2.64
Twelve to Thirteen..	Boys.	1.06	1.67	1.37	1.72	1.74	1.81	2.19	1.77	1.92	2.24	2.19	1.65
	Girls.	3.93	2.98	2.94	3.24	3.07	3.30	3.43	3.21	4.12	3.82	4.39	3.36
Thirteen to Fourteen	Boys.	1.81	1.76	2.31	2.36	2.25	2.22	2.09	2.76	3.23	3.61	4.37	2.64
	Girls.	1.55	2.39	2.64	2.80	2.83	3.09	2.95	3.12	2.46	2.68	2.47	2.65
Fourteen to Fifteen..	Boys.	2.17	2.50	2.08	2.20	2.58	2.96	3.24	3.67	4.04	3.78	3.14	2.90
	Girls.	3.80	3.22	3.38	3.14	2.98	2.59	2.53	2.45	2.25	1.81	1.77	2.74
Fifteen to Sixteen ..	Boys.	2.00	2.09	2.89	2.83	3.86	3.55	3.49	3.08	2.11	2.56	2.88	2.54
	Girls.	3.59	3.72	2.50	2.31	2.27	2.30	2.22	1.96	1.91	1.56	1.85	2.34
Sixteen to Seventeen	Boys.	3.18	3.47	3.74	4.22	3.42	3.00	3.46	3.32	3.14	3.18	2.68	3.40
	Girls.	1.51	0.85	0.52	1.16	1.12	0.91	0.96	0.90	0.67	0.99	0.75	0.90
Seventeen to Eighteen	Boys.		4.26	2.93	2.10	1.66	1.48	1.57	2.47	3.20	1.66	0.65	2.55
	Girls.	1.03	0.82	1.59	0.93	0.71	0.48	0.31	0.28	0.24		0.55	0.54

TABLE No. 34.

THE ABSOLUTE ANNUAL INCREASE IN THE GIRTH OF CHEST; DERIVED FROM TABLE NO. 23.

Age at Nearest Birthday.	Sex.	Value in Centimetres at the following Percentile Grades.											Average of Increase.
		5	10	20	30	40	50	60	70	80	90	95	Centimetr's
Six to Seven......	Boys.	1.89	1.45	1.39	1.28	1.67	1.57	1.69	1.76	1.89	1.87	1.84	1.57
	Girls.	0.43		0.74	0.95	0.86	1.00	1.15	1.28	1.37	1.42	1.32	1.13
Seven to Eight.....	Boys.	1.21	1.38	1.69	1.71	2.06	1.66	1.92	1.95	1.81	1.56	0.87	1.56
	Girls.	2.94	1.27	1.32	1.25	1.83	1.47	1.57	1.53	1.45	1.43	1.03	1.34
Eight to Nine......	Boys.	1.87	1.56	1.42	1.55	1.09	1.78	1.58	1.69	1.88	2.02	2.47	1.72
	Girls.	1.68	1.49	1.69	1.74	1.76	1.69	1.62	1.80	1.80	1.90	2.22	1.70
Nine to Ten.......	Boys.	1.12	1.42	1.46	1.57	1.51	1.59	1.67	1.65	1.80	2.28	2.47	1.69
	Girls.	0.23	0.51	0.50	0.56	0.56	1.43	1.50	1.50	1.63	1.59	1.95	0.51
Ten to Eleven......	Boys.	1.36	1.44	1.62	1.98	1.76	1.76	1.61	1.64	1.73	1.66	1.35	1.65
	Girls.	2.16	2.11	2.12	2.28	2.53	1.90	2.08	2.03	1.89	2.44	2.49	2.83
Eleven to Twelve...	Boys.	1.32	1.31	1.40	1.09	1.44	1.49	1.77	1.75	1.73	1.41	1.64	1.52
	Girls.	1.50	2.07	2.37	2.20	2.13	2.13	2.24	2.63	3.21	3.24	3.36	2.49
Twelve to Thirteen.	Boys.	1.16	1.34	1.69	1.42	1.69	1.84	1.88	1.92	2.11	2.61	2.88	1.85
	Girls.	1.75	2.16	2.53	2.78	2.93	3.15	3.13	3.11	3.06	3.41	3.59	2.95
Thirteen to Fourteen	Boys.	2.27	2.08	1.94	2.36	2.34	2.37	2.69	3.07	3.21	3.33	3.63	2.66
	Girls.	3.08	2.64	2.57	2.85	2.88	2.95	3.16	3.18	3.14	2.96	3.04	2.84
Fourteen to Fifteen.	Boys.	2.22	2.72	2.98	3.04	3.28	3.56	3.41	3.41	3.70	3.77	3.83	3.29
	Girls.	2.36	2.77	2.74	2.53	2.57	2.33	2.42	2.84	2.88	2.60	2.61	2.65
Fifteen to Sixteen..	Boys.	1.13	1.48	2.15	2.58	2.97	2.96	3.06	3.10	2.68	3.31	3.54	2.66
	Girls.	3.24	2.53	2.23	2.40	2.32	2.30	1.97	1.72	1.69	2.36	1.93	2.07
Sixteen to Seventeen	Boys.	1.30	3.84	3.42	3.16	2.65	2.30	1.96	2.10	2.01	1.00	0.03	2.17
	Girls.	1.37	1.56	1.66	1.63	1.37	1.50	1.56	1.19	1.37	1.94	1.26	1.54
Seventeen to Eighteen	Boys.	5.23	4.19	3.62	3.96	3.39	3.15	3.08	2.39	1.84	1.78	1.29	3.13
	Girls.	0.30	0.60	0.18	0.13	0.37	0.20		1.70				0.06

during the years 6 and 7 than at any other time, and in girls is nearly as great in these years as in the period of acceleration. The curve of girls' height sitting is very similar to that of girls' height standing. In both, the rate of growth is more uniform than in weight. Indeed, the period of acceleration in the last named dimension is greater than in any of the others. The relative annual increase of boys' height sitting seems almost atypical, by reason of its sharp ascent at ages 8 and 9 and its failure to sink after age 16. The latter feature is perhaps due to an error in the average value caused by the small number of observations at age 17. I am unable to explain the ascent at ages 9 and 10. The growth in span of arms is somewhat more rapid at ages 7 and 8 than during the prepubertal acceleration. The curve of girls' girth of chest differs from the usual type in its sudden rise at age 11, the increase at that year appearing slightly greater than at age 13. The boys' curve, on the contrary, agrees very well with the curves of weight, height, etc., except that the curve rises at age 18, where the number of observations, it may be repeated, is perhaps too small for very sure work. The quickness with which the strength of squeeze decreases after age 7 is certainly remarkable, as is the sharpness with which the prepubertal acceleration is shown.

Attention will be called in chapter IX, to the importance in children of the relation between height and weight, girth of chest and other physical dimensions. Unusual height, it will be pointed out, is commonly a disadvantage, because it entails an unusual loss of energy. If such individuals have a weight and girth of chest, etc., so much above the common as to compensate their excessive height, they are likely to be able to meet all demands on their strength. If they do not possess this compensatory development, they will probably be unable to meet any excessive demand. Thus the question of how far this compensation exists in any individual, or, more broadly stated, the question what weight, girth of chest, etc., should accompany any given height, becomes of the greatest interest. This interest, it should be remarked, is unusually great in the case of children, for children are taxed with the mechanical motion and other forms of dissipation of energy making up the ordinary output of life, and, in addition, with the extraordinary function of storing energy in the increase of tissue which constitutes growth.

In view of these facts, Plate XLV cannot fail to be instructive. It shows the ratio of span of arms, height sitting, chest-girth, weight, strength of squeeze with right hand, and five head and face measurements to height standing. Height standing is here expressed by an abscissa, and the percentage relation of weight and the other dimensions are displayed in curves. Of all these, span of arms most closely approximates the height, the difference being less than one per cent. of the latter from age 6 to 11 and scarcely more than 2 per cent. in subsequent ages. Span of arms in both sexes is therefore nearly the same as height standing throughout the period of observation, becoming very slightly greater than the height as growth progresses. The height sitting and the girth of chest run a parallel course and are, moreover, nearly equal, the girth of chest being about 2 per cent. less than the height sitting. They increase a little less rapidly than the height, showing a decline of about 4 per cent. in thirteen years. Height sitting and chest-girth are not far from half the height standing.

Far different is the development of weight and strength of squeeze. These increase much more rapidly than height, for at age 6 the height stands to weight in the ratio of 100 to 18 and to strength of squeeze as 100 to 6, while at age 16 these ratios are 100 to 34 and 100 to about 16, respectively. The parallelism in the development of weight and strength of squeeze is of much interest. The dimensions of head and face increase somewhat less rapidly than the height. The length of head, for example, falls from $\frac{1}{16}$th of the height, at age 6, to about $\frac{1}{17}$th at age 18.

CHAPTER VIII.

THE RELATION BETWEEN THE PHYSICAL DEVELOPMENT OF SCHOOL CHILDREN AND THEIR CAPACITY FOR MENTAL LABOR.

In Vol. VI, No. 7 of the Transactions of the Academy of Science of St. Louis, issued March 21, 1893, I demonstrated that children who possess more than the ordinary power of mental labor, as measured by their progress in their studies, are heavier, taller and larger in girth of chest and in width of head than their less gifted companions of the same age. A more extended statement of these observations was presented to the Berliner Gesellschaft für Anthropologie, Ethnologie und Urgeschichte, July 15, 1893, and appears in Virchow's Zeitschrift für Ethnologie under the title Untersuchungen der Schulkinder in Bezug auf die physischen Grundlagen ihrer geistigen Entwickelung. In these papers, the material was the total number of observations irrespective of the social condition of parents. An example, selected from Tables Nos. 2 and 4, page 165 and 167, of The Physical Basis of Precocity and Dullness, will illustrate the result of the inquiry. Pupils aged 11 are found in Grades I, II, III, IV, V and VI of the St. Louis Public Schools, as the following table shows. The 59 boys of Grade I, the lowest grade,

TABLE No. 35.

MEDIAN WEIGHT OF BOYS AGED 11 DISTRIBUTED BY SCHOOL GRADE.

Grades.	No. of Boys Weighed.	Median Weight.
I	59	28.83 Kg.
II	311	29.74 "
III	664	30.92 "
IV	546	31.43 "
V	123	32.41 "
VI	33	33.29 "

weigh less than the boys of Grade II, and these, again, are lighter than the boys in higher grades.

It is a well-known fact that children of the prosperous classes are better developed physically than the children of the poor. It has been suggested that the children of poor parents are not so successful in school work as the children of the rich, and that the poor children are compelled to leave school at an earlier age than the rich, and for these reasons are relatively less numerous than the rich in the higher grades. According to this idea, the better physical development of the children of the same age in the higher grades, as illustrated above in boys aged 11, is due to the preponderance in the higher grades of the well nourished children of the rich. This may be, and probably is, a factor in the phenomenon, but is surely only a partial influence.

For when children of the same age and as nearly as possible of the same social class are weighed and the weights distributed by school grade, it is found that they follow the law established for the whole material irrespective of social condition. A glance at Plate XLVI, derived from Table No. 36, will convince the reader of the truth of this statement.

TABLE No. 36.

MEDIAN WEIGHT OF THE DAUGHTERS OF MANUAL TRADESMEN DISTRIBUTED BY SCHOOL GRADE.

Age at Nearest Birthday.	Unit of Measurement and Number of Observations.	Kindergarten.	SCHOOL GRADES.						
			I.	II.	III.	IV.	V.	VI.	VII.
Seven.......	Kilogr. No. of Obs.	19.73 137	21.14 187						
Eight.......	Kilogr. No. of Obs.		22.60 287	23.56 136					
Nine........	Kilogr. No. of Obs.		24.00 87	25.33 240	25.79 68				
Ten.........	Kilogr. No. of Obs.			27.03 152	27.87 170	28.71 33			
Eleven......	Kilogr. No. of Obs.			27.97 49	29.22 135	30.11 116	29.74 27		
Twelve......	Kilogr. No. of Obs.				31.95 65	32.57 140	33.69 76	34.50 32	
Thirteen.....	Kilogr. No. of Obs.				34.16 32	36.32 78	36.66 69	39.16 56	40.18 27
Fourteen.....	Kilogr. No. of Obs.					38.59 25	41.31 40	40.63 60	

Further evidence is afforded by Table No. 37, in which the daughters of professional men are divided into two equal groups, in the manner explained on page 177 of The Physical Basis of Precocity and Dullness, one group containing the fifty per cent who are found in the upper grades, the other the fifty per cent found in the lower grades. It should be

TABLE No. 37.

MEDIAN WEIGHT OF THE DAUGHTERS OF PROFESSIONAL MEN DISTRIBUTED BY SCHOOL GRADE.

Age at Nearest Birthday.	Number of Observations.	School Grades.	Median Weight. (Kilogram's)	Median Weight. (Kilogram's)	School Grades.	Number of Observations.	Age at Nearest Birthday.
7	50	Kg., 10 I.	20.29	20.81	90 I, II.	50	7
8	52.5	Kg., 97 I.	22.13	23.75	3 I, II, III.	52.5	8
9	53.5	I, 69 II.	24.52	25.07	31 II, III, IV, V.	53.5	9
10	57.5	I, II, 48 III.	27.19	27.64	52 III, IV, V, VI.	57.5	10
11	58.0	I, II, III, 28 IV.	28.76	31.27	72 IV, V, VI, VII.	58.0	11
12	50	I, II, III, 81 IV.	33.65	34.05	19 IV, V, VI, VII.	50	12
13	55.5	II, III, IV, 78 V.	38.46	39.55	22 V, VI, VII, VIII.	55.5	13

remarked that this method of division diminishes the conspicuousness of the difference between lower and higher grades by not presenting the weights for outlying grades, but for a small number of observations it is much more reliable than the method of which Table No. 36 is an illustration.

The results of this study of the weights of girls from the same social class distributed by school grade confirm the conclusion reached in the publications cited above, namely that successful pupils are larger than the unsuccessful.

CHAPTER IX.

THE APPLICATION TO INDIVIDUAL SCHOOL CHILDREN OF THE MEAN VALUES DERIVED FROM ANTHROPOLOGICAL MEASUREMENTS BY THE GENERALIZING METHOD.

The data for the studies described in this work can be collected either by the "generalizing" or by the "individualizing" plan. In the former, a great number of measurements is made but once on individuals of different ages, and the measurements classified according to age. In the latter, the same individuals are measured yearly or oftener during their period of growth, and the measurements classified also by age. The generalizing method, the one pursued in the present investigation, is rapidly and easily carried out, whereas the individualizing method demands for its execution exceptional opportunities and exceptional patience, requiring not only that the measurements be made and the records kept through two decades, but that the number of children measured in the early years of this long period be very great, lest death and desertion so thin the ranks that those remaining to the end shall be too few to yield trustworthy conclusions. Both methods, when applied to the same material, give identical results with regard to means, including those of subdivisions as well as those of the whole number of observations. The individualizing method does more.

The importance of the individualizing method has been much emphasized, for the reason that it can give information without which the laws derived from means cannot, in the present state of knowledge, be applied to individuals. Before this application can be made, it is necessary to know the degree of probability that an individual, who at a given age stands at a certain deviation from the mean of any dimension will show the same deviation at other ages; for example, the degree of probability that a girl whose height at age 8 is 122.06 cm., and who therefore deviates 3.7 cm., or $+1d$ from the mean of her age (118.36 cm.), will deviate to the same degree ($+1d$) from the mean height throughout her growth. In that case, the law of growth for the type at a deviation of $+1d$ from the mean is her law of growth. Otherwise, she is an exception and practical regulations deduced from the law for the type cannot be safely made binding on her. This knowledge, as has just been said, is furnished by the individualizing method, while the generalizing method is of no assistance in this matter.

The application to individuals of the law of growth of the mean is a subject of immediate practical interest. The connection between theory and practical affairs is here unusually short and clear. Were this application possible, the deviations of children from the laws of normal growth could be quickly recognized and largely overcome, the evil effect of over study could be watched and intelligently combated, and systems of education, no longer exacting of all that which should be exacted only from the mean, could be rationally adapted to the special needs of the exceptionally weak and the exceptionally strong.

These beneficent reforms, it is at present believed, must await the slow collection of data by the individualizing method. If it is indeed true that the laws of growth determined for the mean cannot be used for the individual until the individualizing method has established the probability of each individual deviation remaining constant throughout the period of growth, then a generation must elapse — so slow is the gathering of data by this method — before the necessary knowledge is in our hands. I hope to show that this long waiting is unnecessary, and that the data collected by the generalizing method may be used, in a way hitherto unrecognized, for the making of standards by which the deviation of an individual from the mean of his age may be seen to be normal or abnormal.

Let the problem be clearly understood. The question is: this boy or girl is above or below the mean height, or weight, etc., of his or her age. How shall it be known that this deviation is normal or abnormal? There has been hitherto no satisfactory reply to this question. A vague and partial answer is possible after two measurements separated by at least a year's interval. If the deviation is the same, or very nearly the same, at both measurements, the probability is that the child is growing normally. This probability is greater than the general probability that a normal deviation is more likely to occur than an abnormal one, but its numerical value is wholly unknown. If, on the other hand, the two deviations are unequal, the probability is that the greater of them is abnormal, but the numerical value is here also unknown. How definitely the individualizing method could answer this question is difficult of conjecture, in the present lack of data, but certainly no answer whatever could be expected except after two measurements separated by a year's interval, a year in which the unchecked cause of an abnormal deviation might inflict an irreparable damage on the organism. Such indefinite and fragmentary knowledge can never be the basis of a practical reform. Any solution of this problem which shall gain general acceptance must be easy to understand and easy to apply, and must give the probable degree of abnormality of any observed deviation. These conditions are, I believe, fulfilled by the following method.

According to the theory of probabilities, the heights of a thousand individuals of the same class will arrange themselves as follows: —

Between	$M+4d$	and	$M+nd$	3	individuals.
"	$M+3d$	"	$M+4d$	18	"
"	$M+2d$	"	$M+3d$	67	"
"	$M+ d$	"	$M+2d$	162	"
"	M	"	$M+ d$	250	"
"	M	"	$M- d$	250	"
"	$M- d$	"	$M-2d$	162	"
"	$M-2d$	"	$M-3d$	67	"
"	$M-3d$	"	$M-4d$	18	"
"	$M-4d$	"	$M-nd$	3	"

where $M =$ the mean and $d =$ the probable deviation.

Let these be divided into seven groups: —

I.	All individuals between	$M+nd$	and	$M+3d$	21
II.	"	"	"	$M+3d$	"	$M+2d$	67
III.	"	"	"	$M+2d$	"	$M+ d$	162
VI.	"	"	"	M	"	$M\pm d$	500
V.	"	"	"	$M- d$	"	$M-2d$	162
IV.	"	"	"	$M-2d$	"	$M-3d$	67
VII.	"	"	"	$M-3d$	"	$M-nd$	21

The mean height, weight, girth of chest, etc., of each of these groups at any age will be the type of a certain degree of deviation from the mean of the age. That is to say, the weights, etc., of each group will be symmetrically distributed above and below the mean weight, etc., of the group in the manner already illustrated for the entire undivided number of observations, i. e., the entire thousand. Each group, therefore, will be characterized by a physical development definitely determined by the means of height, weight and other physical dimensions. These means, taken together form the type or norm of the group. The individual deviations from this norm follow the theory of probability, and the degree of abnormality presented by any individual deviation can be expressed in the terms of this theory. An example will illustrate this. A boy X shows a deviation in height of $+1.5d$ from the mean height of his age. He falls therefore in Group III. The boys in this group possess a mean weight of M kilogrammes, with a probable deviation of $\pm d$; that is, boys between d and $2d$ taller than the norm of their age should weigh $M \pm d$ kilogrammes. In like manner, they should have a girth of chest of $M \pm d$ centimetres, and a span of arms of $M \pm d$ centimetres, and so on. If the weight, etc., of the boy X coincide with the means of his group (Group III) his physique is normal, the accuracy of this conclusion being proportionate to the number of different dimensions on which it is based. If the boy X deviate more than $\pm 1d$ from the mean in one or more dimensions, his development is abnormal, and the degree of abnormality is measured by the amount of his deviation.

The necessity of choosing some one dimension as the basis of such a system is self-evident. There are good reasons, partly theoretical and partly practical, why height rather than weight should be taken as a basis. Height is more stable, less liable to irrelevant fluctuations, than weight. An excess in weight can be reduced; a child whose weight is out of proportion to its height may be brought into proportion by suitable diet and exercise; but height, once attained, cannot be reduced, nor can the growth in height be easily influenced. Practically, therefore, the physical trainer must be content to bring the weight, girth of chest, strength of squeeze and other physical dimensions up to the mean development which corresponds to the height of the child. Experience has abundantly shown that the relation of weight to height is of great importance to health, life insurance companies declining to receive applicants whose weight falls much below the standard weight of their height. For these reasons, height should be preferred as the basis of the system.

The question whether any given deviation is normal or abnormal is answered by this system in two ways: in respect of height, by the degree of deviation from the mean or norm of the whole number of observations; in respect of other dimensions, by the degree of deviation of the weight, girth of chest, etc., from the mean weight or girth of chest corresponding to the height of the individual under examination, this normal weight, etc., being determined with sufficient exactness by taking the means and probable deviations of the group in which the height falls. It is evident that all cases included within $M \pm d$ must be termed normal, for the chances are even that any individual measurement in a series will fall within $M \pm d$ and are against its exceeding these limits, being 4.64 against 1 that it will fall at $M \pm 2d$.

Strictly speaking, all abnormal deviations in any dimension are probably unhealthful, yet an important difference exists in this respect between abnormal deviations in height and abnormal deviations in weight, girth of chest, etc., as related to height. It cannot be doubted that abnormal height is probably (using the word in its technical sense) a disadvantage. The potential energy of the body is converted into mechanical labor and heat, by far the greater expenditure taking the latter form. In the adult, the total expenditure in the form of heat is about 2,700 calories in 24 hours (Helmholtz), of which 80.1 per cent escape in radiation, conduction and evaporation from the skin. Thus the superficies of the body plays a great part in the dissipation of energy. The superficies is greater, usually, in tall children than in short, a difference of special importance in the young, in whom metabolism is much more active than in the adult. More heat is therefore lost by the abnormally tall than by those of normal height. There is a disadvantage also in the loss by mechanical labor. Greater height entails increased work on the heart and on the skeletal muscles. In short, increased loss of energy goes hand in hand with increase in height. Hence in the tall the necessity for a physical development which shall be so much above the mean as to compensate their greater loss of energy. In growing children not only must there be compensation for the expenditure of energy, but there must be also energy stored in the increase of tissue which constitutes growth.

If the greater demands of tall children are balanced by a correspondingly greater income of energy, a normal equilibrium or "health" is preserved. It should be clearly recognized that this equilibrium is unaffected by the absolute height and is dependent only on the relation between height and the other physical dimensions. Consequently health is as possible in tall children as in those of normal height, although less probable, for the chances against a compensatory development of weight and other dimensions increase very rapidly with the deviation of the height from the norm. The absolute height of an individual is of very secondary interest from a practical point of view, because it is not necessarily a state of ill health, whereas the development of weight, girth of chest, etc., in proportion to height is of supreme interest. A lack of proportion between height and other physical dimensions is itself ill health. The tendency of organisms to adopt ends to means is strong, and an imperfect compensation may suffice for most demands. A heart in which an hypertrophy of the left ventricle has partially compensated an insufficiency of the mitral valve may beat regularly enough for ordinary exertions, and yet fail utterly when its possessor is forced suddenly to ascend a height or to make any other unusual exertion. So a tall child may have a sufficient income of energy to meet the demands of a wisely regulated life, and sink under the burden of unusual tasks.

It has been shown in the foregoing pages that the means derived from anthropometrical metrical measurements by the generalizing method can be used to determine whether the weight and other physical dimensions of an individual are normal in relation to height, and it has been pointed out that this normal relation constitutes health. It follows that the normal amount of labor cannot be exacted without injury from those in whom this equilibrium is wanting. These facts must therefore be taken into account in a rational school system, and it should now be made plain how this is to be done.

All systems of education have for their object the largest possible development of individual minds. In large schools the tasks by which this development is promoted are those which secure from the child of mean ability its maximum mental output. In practice they are determined by examinations. Hence the existence in every educational institution of classes based on the mental output of the mean pupil, and related to age only in that the output fixed as the standard of any class is necessarily found more often at a certain age than at other ages. Thus there exists a mean age for each class; the greater number of pupils at any age is found in the same class, while some have advanced beyond, and others, equally old, have not yet come so far as this class. On an average, those who have advanced beyond the greater number of their age are precocious, that is, possess more than the mean capacity for mental labor, while those who are less advanced are dull, possessing less than the mean capacity. It has been demonstrated that there is a physical basis for precocity and dullness:[1] when numbers sufficiently large for statistical purposes are employed, it is seen that precocious pupils are of greater mean weight, height, etc., than the mean pupils and that the latter are heavier and taller that the dull. The mental output is therefore directly related to the physical condition of the pupils. The mean height, weight, girth of chest, etc., in any grade is the mean physical development corresponding to the mental output of the grade. It follows that those who do not possess this development cannot without abnormal strain do the work exacted in this grade. On the other hand, pupils who possess more than the mean

[1] See Chapter VIII, page 335 et seq.

physical development of their age should be capable of more than the mean labor. Yet the management of this latter class presents but few difficulties, whereas the former class cannot be too carefully protected.

The consequences of continued overstrain in a growing boy or girl are most unhappy. The curves of growth in height and weight of the mean child are characteristic. The quick rise to age 7 or 8, the slower ascent to age 11 in girls and 13 in boys, the remarkable three years of accelerated development preceding puberty, and, finally, the rapid decrease in the rate of growth as full development approaches express the normal development of the type and, presumably, the normal development of the individual. Overwork may cause a temporary or a permanent deviation in these curves. It is probable, though not certain, that a temporary loss consequent on a slight overstrain may not lower the final outcome of the development, but there can be no doubt as to the result of a prolonged strain. In such a case, the probability is strong that the whole subsequent curve will be turned out of its course. A prolonged strain in a growing child harms for life and leaves a mark which can never be effaced. The danger is greatest in the periods of quickest development, particularly great in the prepubertal period. It is a sufficient commentary on the evils of the present educational methods that during these very years the undiscriminating routine of a system devised for the average pupil is most inflexibly applied to weak and strong alike.

Overstrain can often be recognized both by subjective and objective symptoms. Subjective symptoms, however, are frequently obtained with difficulty, especially in pupils who are unusually ambitious and who overstudy from choice. An objective symptom must therefore be found — a symptom easily demonstrated and almost never wanting. Such a symptom is the failure to gain weight at the normal rate. A persistent loss of weight in an adult is regarded as a matter of grave concern: the persistent failure of a child to make the normal gain in weight is no less grave. It is not pretended that the failure to gain weight always accompanies overstrain, but it is claimed that the number of exceptions is small and that frequent weighing is the most practical and on the whole the most certain method of detecting the presence of influences that are working injury to the development of the child. The skillful breeder of cattle depends on systematic weighing to inform him if his efforts are meeting with success, but children are left to grow at haphazard.

It is not enough that overstrain should be recognized by the harm it has done. The child should be guarded against the possibility of harm. The anthropometrical system proposed offers a means of doing this. It infallibly discovers the children whose physical development is below the standard of their age. It no less certainly indicates the physical development which most often accompanies the power to do the mental work of any grade. It therefore divides the pupils into two bodies; those physically competent and those physically incompetent for a clearly defined degree of mental exertion. When working with great numbers, the infallibility of this system is practically absolute and theoretically almost absolute. When applied to individuals, errors will certainly occur, but the number of errors will according to the laws of probability be less than the number of correct conclusions, and these errors cannot influence the great fact that such a system is competent to call attention to the children who will probably be unable to do the normal work of their age without injury. Each individual case must then be treated on its own merits.

The proposed system of physical examination requires: —

I. The collection of sufficiently extensive data by the generalizing method.

II. The determination of the means and the probable deviations of height, weight, girth of chest, strength of squeeze, etc., for each age.

III. The division of the individuals at each age into groups in terms of the probable deviation from the mean height, as illustrated above, and the calculation of the mean and probable deviation of the weight, girth of chest, etc., of each group.

IV. The determination of the mean physical development of the pupils in each class or grade of the school system.

V. The physical examination of each applicant for entrance to any grade.

These data permit the enforcement of the following regulation: That no pupil whose physical development varies more than $\pm d$ from the weight, etc., of the mean pupil of his height in a class which his mental output would otherwise entitle him to enter, shall be admitted to that class, unless with the approval of a medical expert, if possible a regularly appointed school physician, who shall testify that the pupil's strength will be equal to the strain.

TABLES NO. 38 TO 51.

The tables which follow this page repeat some of the more commonly used statistical values in a form which admits of ready reference and comparison. Such are the average, probable error of the average, probable deviation, median or 50 percentile grade and median minus average values. Yet these tables are by no means summaries of the statistical calculations of the investigation, since they omit much that is given in the foregoing pages. They contain, moreover, not a little new matter. The relation of probable deviation to average, relation of average to height standing, relative annual increase of average, the 25 percentile grade, the 75 percentile grade, the cranial indices and the absolute annual increase of average of strength of squeeze and of measurements of the head and face are here presented for the first time.

Some of this material has been already discussed in this paper. Some is reserved for future discussion. One series, the relation of probable deviation to average, was treated at some length in my paper on "The relation between the growth of children and their deviation from the physical type of their sex and age," *Transactions of the Academy of Science of St. Louis*, Vol. VI, No. 10, pp. 233–250, November 14, 1893, to which the reader is referred.

Five dynamometers were used in testing the strength of squeeze; they were distinguished by the first five letters of the alphabet; A and D were graduated alike; C and E were also alike but differed from A and D, and B differed from all the others. The original values obtained with these various instruments are set down in Tables No. 45 and 46; from them were made Tables No. 43 and 44, by reducing the arbitrary graduation of all dynamometers to a uniform scale in kilogrammes.

350

TABLE No. 38.
The Height Standing.

AGE AT NEAREST BIRTHDAY.	SEX.	Number of Observations.	Unit of Measurement.	Average.	Probable Error of Average. E.	Probable Deviation. d.	Relation of Probable Deviation to A'v'age. d/A	Absolute Annual Increase of Average.	Relative Annual Increase of Average. %	25 Percentile Grade.	Median or 50 Percentile Grade.	75 Percentile Grade.	Median Minus Average.
Six	Boys.	709	Centimetr's	108.94	±0.128	±3.40	3.1			105.99	109.23	112.69	+0.29
	Girls.	780	"	107.67	0.123	3.42	3.2			104.74	108.10	111.30	+0.43
Seven	Boys.	1850	"	114.03	0.084	3.61	3.2	5.08	4.7	111.02	114.48	118.02	+0.45
	Girls.	1791	"	112.95	0.089	3.75	3.3	5.28	4.9	109.72	113.44	117.07	+0.49
Eight	Boys.	2223	"	119.13	0.082	3.89	3.3	5.10	4.7	116.08	119.77	123.58	+0.64
	Girls.	2193	"	118.36	0.079	3.70	3.1	5.41	4.8	115.07	118.74	122.35	+0.38
Nine	Boys.	2205	"	124.35	0.080	3.75	3.0	5.22	4.4	121.09	124.87	128.76	+0.52
	Girls.	2122	"	123.67	0.083	3.83	3.1	5.31	4.5	120.90	124.11	127.58	+0.44
Ten	Boys.	2087	"	128.87	0.087	3.98	3.1	4.52	3.6	125.49	129.45	133.54	+0.68
	Girls.	2053	"	128.43	0.089	4.06	3.1	4.76	3.9	124.96	128.85	132.58	+0.42
Eleven	Boys.	1819	"	133.84	0.099	4.23	3.2	4.97	3.9	130.03	134.44	138.29	+0.60
	Girls.	1772	"	133.19	0.106	4.48	3.2	4.76	3.7	129.23	133.60	137.97	+0.41
Twelve	Boys.	1653	"	138.21	0.116	4.47	3.2	4.37	3.3	134.25	138.57	142.99	+0.36
	Girls.	1732	"	139.11	0.098	5.23	3.4	5.92	4.4	134.61	139.54	144.44	+0.43
Thirteen	Boys.	1268	"	142.91	0.140	4.98	3.5	4.70	3.4	138.45	143.29	148.28	+0.38
	Girls.	1322	"	146.53	0.150	5.46	3.8	7.42	5.3	140.98	146.19	151.79	−0.34
Fourteen	Boys.	925	"	146.58	0.183	5.58	3.8	5.67	4.0	143.43	148.86	154.52	+0.28
	Girls.	1085	"	150.84	0.156	5.15	3.7	4.31	2.9	146.86	151.94	156.32	+1.10
Fifteen	Boys.	490	"	154.90	0.286	6.33	4.1	6.32	4.3	148.88	155.25	161.86	+0.35
	Girls.	680	"	155.04	0.154	4.01	3.4	4.20	2.8	151.75	155.82	159.58	+0.78
Sixteen	Boys.	189	"	160.27	0.427	5.87	3.7	5.37	3.5	155.28	161.27	167.04	+1.00
	Girls.	420	"	157.52	0.197	4.05	2.6	2.48	1.6	153.94	158.03	161.81	+0.51
Seventeen	Boys.	78	"	165.13	0.592	5.15	3.1	4.86	3.0	160.79	166.00	171.34	+0.87
	Girls.	206	"	159.33	0.241	3.45	2.2	1.81	1.1	155.65	159.40	162.93	+0.07
Eighteen	Boys.	29	"	170.41	0.924	4.98	2.8	5.28	3.2		170.50		+0.09
	Girls.	164	"	159.42	0.265	3.39	2.1	0.09	0.06	156.40	159.50		+0.08
Nineteen	Girls.	85	"	158.46	0.438	4.04	2.6			154.32	159.56	163.29	+1 10
Twenty	Girls.	79	"	159.41	0.353	3.08	2.0				159.83		+0.42
Twenty-one	Girls.	43	"	159.98	0.651	4.27	2.7				160.50		+0.52

TABLE No. 39.

THE WEIGHT.

AGE AT NEAREST BIRTHDAY.	SEX.	Number of Observations.	Unit of Measurement.	Average.	Probable Error of Average. E	Probable Deviation. d	Relation of Probable Deviation to Av'age. $\frac{d}{\text{Av}}$ %	Relation of Av'age to Height Standing. %	Absolute Annual Increase of Av'age.	Relative Annual Increase of Av'age. %	25 Percentile Grade.	Median or 50 Percentile Grade.	75 Percentile Grade.	Median Minus Average.
Six	Boys.	707	Kilogram.	19.75	±0.054	±1.43	7.2	18.1			18.57	19.86	21.27	+0.10
	Girls.	798	"	18.93	0.051	1.44	7.6	17.6			17.70	18.99	20.44	+0.06
Seven	Boys.	1814	"	21.67	0.039	1.68	7.8	19.0	1.92	9.7	20.14	21.66	23.36	−0.01
	Girls.	1714	"	20.82	0.045	1.88	9.0	18.4	1.89	10.0	19.25	20.81	22.55	−0.01
Eight	Boys.	2188	"	23.78	0.042	1.96	8.2	20.0	2.11	9.7	22.00	23.87	25.86	+0.09
	Girls.	2147	"	22.88	0.042	1.95	8.5	19.3	2.06	9.9	21.02	22.85	24.75	−0.03
Nine	Boys.	2188	"	26.06	0.045	2.09	8.0	21.0	2.28	9.6	24.17	26.22	28.28	+0.16
	Girls.	2055	"	25.08	0.049	2.23	8.9	20.3	2.20	9.6	23.07	25.04	27.31	−0.04
Ten	Boys.	2064	"	28.32	0.049	2.23	7.9	22.0	2.26	8.7	26.07	28.36	30.81	+0.04
	Girls.	1947	"	27.49	0.062	2.31	8.4	21.4	2.41	9.6	25.08	27.45	29.96	−0.04
Eleven	Boys.	1743	"	31.00	0.062	2.60	8.4	23.2	2.68	9.5	28.50	31.08	33.66	+0.08
	Girls.	1708	"	30.15	0.070	2.91	9.6	22.6	2.66	9.7	27.28	29.80	32.84	−0.35
Twelve	Boys.	1644	"	33.51	0.061	2.46	7.3	24.2	2.51	8.1	30.65	33.41	36.26	−0.10
	Girls.	1676	"	33.66	0.081	3.31	9.8	24.2	3.51	11.6	30.05	33.25	36.83	−0.41
Thirteen	Boys.	1242	"	36.61	0.110	2.88	10.6	25.6	3.10	9.3	32.88	36.25	39.86	−0.36
	Girls.	1343	"	38.49	0.115	4.22	11.0	26.3	4.83	14.3	33.86	38.00	42.79	−0.49
Fourteen	Boys.	946	"	40.44	0.148	4.56	11.3	27.2	3.83	10.5	36.41	39.98	45.38	−0.46
	Girls.	1082	"	42.29	0.142	4.67	11.0	28.0	3.80	9.9	37.87	42.65	47.55	+0.36
Fifteen	Boys.	498	"	46.22	0.227	5.06	11.0	29.8	5.78	14.3	40.04	45.49	51.75	−0.73
	Girls.	690	"	46.69	0.154	4.05	8.7	30.1	4.40	10.4	42.59	46.85	50.66	+0.16
Sixteen	Boys.	203	"	51.60	0.431	6.16	12.0	32.0	5.38	11.6	45.66	51.83	57.99	+0.23
	Girls.	420	"	50.25	0.207	4.24	8.4	31.9	3.56	7.6	46.56	49.96	54.96	−0.29
Seventeen	Boys.	71	"	55.67	0.521	4.38	7.9	33.7	4.07	7.9	51.94	55.31	61.21	−0.36
	Girls.	230	"	52.61	0.244	3.70	7.0	33.0	2.36	4.7	48.67	52.52	56.47	−0.09
Eighteen	Girls.	165	"	52.36	0.289	3.60	6.9	33.0			49.27	53.10	56.83	+0.74
Nineteen	Girls.	81	"	52.19	0.332	3.76	7.2	32.8			48.93	52.47	55.74	+0.28
Twenty	Girls.	66	"	53.91	0.565	3.76	7.0	33.7			50.01	53.57	57.33	−0.34

352   Trans. Acad. Sci. of St. Louis.

TABLE No. 40.
THE HEIGHT SITTING.

AGE AT NEAREST BIRTHDAY.	Sex.	Number of Observations.	Unit of Measurement.	Average.	Probable Error of Average. E.	Probable Deviation d.	Relation of Probable Deviation to Average. $\frac{d}{A}$ %	Relation of Average to Height Standing. %	Absolute Annual Increase of Average.	Relative Annual Increase of Average. %	25 Percentile Grade.	Median or 50 Percentile Grade.	75 Percentile Grade.	Median Minus Average.
Six	Boys. Girls.	714 751	Centimeir's "	61.31 59.45	±0.105 0.182	±2.82 2.03	4.6 3.4	56.3 55.2			59.11 58.23	60.77 60.11	63.02 62.05	−0.54 +0.66
Seven	Boys. Girls.	1853 1727	" "	63.32 61.80	0.061 0.053	2.64 2.19	4.2 3.5	55.5 54.7	2.01 2.45	3.3 4.1	61.16 60.58	63.22 62.44	65.22 64.56	−0.10 +0.64
Eight	Boys. Girls.	2239 2120	" "	64.74 63.97	0.048 0.044	2.26 2.04	3.5 3.2	54.4 54.0	1.42 2.17	2.2 3.5	63.22 62.71	65.21 64.96	67.17 66.37	+0.47 +0.99
Nine	Boys. Girls.	2258 2071	" "	66.73 66.16	0.049 0.046	2.34 2.11	3.5 3.2	53.7 53.5	1.99 2.19	3.1 3.4	65.05 64.78	67.08 66.19	69.32 68.82	+0.35 +0.03
Ten	Boys. Girls.	2118 2037	" "	69.25 68.19	0.040 0.049	2.42 2.19	3.5 3.2	53.7 53.1	2.52 2.03	3.8 3.1	66.90 66.72	69.18 68.88	71.35 70.99	−0.07 +0.69
Eleven	Boys. Girls.	1828 1748	" "	70.67 70.03	0.050 0.057	2.56 2.37	3.6 3.4	52.8 52.6	1.42 1.84	2.1 2.7	68.78 67.96	71.09 70.62	73.23 72.94	+0.42 +0.59
Twelve	Boys. Girls.	1656 1707	" "	72.55 72.67	0.067 0.063	2.72 2.61	3.8 3.6	52.5 52.2	1.88 2.64	2.7 3.8	70.49 70.69	72.68 73.22	75.03 75.78	+0.13 +0.55
Thirteen	Boys. Girls.	1285 1354	" "	74.20 76.03	0.076 0.078	2.74 2.87	3.7 3.8	51.9 51.9	1.65 3.36	2.3 4.6	72.03 73.78	74.49 76.52	76.88 79.44	+0.29 +0.49
Fourteen	Boys. Girls.	936 1065	" "	76.84 78.68	0.103 0.095	3.15 3.11	4.1 4.0	51.7 52.2	2.64 2.65	3.6 3.5	74.37 76.50	76.71 79.61	79.87 82.23	−0.13 +0.93
Fifteen	Boys. Girls.	498 674	" "	79.74 81.42	0.161 0.098	3.59 2.54	4.5 3.1	51.5 52.5	2.90 2.74	3.8 3.5	76.51 79.76	79.67 82.20	83.73 84.58	−0.07 +0.78
Sixteen	Boys. Girls.	193 411	" "	82.28 83.76	0.250 0.116	3.48 2.36	4.2 2.8	51.3 53.2	2.54 2.34	3.2 2.9	79.37 82.16	83.22 84.50	86.32 86.52	+0.94 +0.74
Seventeen	Boys. Girls.	77 202	" "	85.68 84.66	0.427 0.153	3.77 2.17	4.4 2.6	51.9 53.2	3.40 0.90	4.1 1.1	83.35 83.00	86.62 85.41	89.55 87.30	−0.94 +0.75
Eighteen	Boys. Girls.	31 167	" "	88.23 85.20	0.536 0.133	2.89 1.72	3.3 2.0	51.7 53.7	2.55 0.54	3.1 0.6	85.86 84.26	87.70 85.89	92.39 87.56	−0.53 +0.69
Nineteen	Girls.	85	"	84.86	0.198	1.82	2.1	53.3			83.96	85.68	87.27	+0.82
Twenty	Girls.	78	"	85.31	0.230	2.03	2.4	53.3			84.15	86.00	88.08	+0.69
Twenty-one	Girls.	41	"	85.05	0.290	1.86	2.2					85.79		+0.74

TABLE No. 41.
THE SPAN OF ARMS.

AGE AT NEAREST BIRTHDAY.	Sex.	Number of Observations.	Unit of Measurement.	Average.	Probable Error of Average. E.	Probable Deviation. d.	Relation of Probable Deviation to Average. d/A	Relation of Average to Height Standing.	Absolute Annual Increase of Average.	Relative Annual Increase of Average.	25 Percentile Grade.	Median or 50 Percentile Grade.	75 Percentile Grade.	Median Minus Average.
					±	±	%	%		%				
Six	Boys.	708	Centimetre.	108.95	1.044	3.85	3.5	100.0			105.66	109.57	113.60	+0.62
	Girls.	769	"	106.96	0.140	3.87	3.6	99.4			104.18	107.29	111.41	+0.33
Seven	Boys.	1862	"	114.42	0.096	4.16	3.6	100.3	5.47	5.0	111.09	115.08	118.87	+0.66
	Girls.	1724	"	112.36	0.107	4.18	3.7	99.5	5.40	5.0	108.76	113.11	117.27	+0.75
Eight	Boys.	2234	"	120.07	0.088	4.18	3.5	100.8	5.65	4.9	116.52	120.73	124.71	+0.66
	Girls.	2152	"	118.33	0.092	4.28	3.6	100.0	5.97	5.3	114.70	118.13	122.53	−0.20
Nine	Boys.	2272	"	125.18	0.089	4.25	3.4	100.7	5.11	4.3	121.55	125.67	129.93	+0.49
	Girls.	2065	"	123.63	0.092	4.18	3.4	100.0	5.30	4.5	120.15	124.28	128.42	+0.65
Ten	Boys.	2076	"	130.22	0.103	4.70	3.6	101.0	5.04	4.0	126.19	130.61	135.49	+0.39
	Girls.	2045	"	128.75	0.104	4.69	3.6	100.2	5.12	4.1	124.63	129.18	133.74	+0.43
Eleven	Boys.	1810	"	135.13	0.113	4.84	3.6	100.9	4.91	3.8	130.89	135.59	140.57	+0.46
	Girls.	1757	"	134.24	0.116	4.87	3.6	100.8	5.49	4.3	129.42	134.56	139.48	+0.22
Twelve	Boys.	1664	"	140.60	0.112	4.57	3.2	101.7	5.47	4.1	135.55	140.48	145.43	−0.12
	Girls.	1718	"	140.07	0.109	4.51	3.2	100.7	5.83	4.3	135.12	140.60	146.05	+0.53
Thirteen	Boys.	1281	"	145.09	0.159	5.71	3.9	101.5	4.49	3.2	140.08	145.33	151.37	+0.24
	Girls.	1368	"	147.19	0.150	5.55	3.8	100.5	7.12	5.1	142.10	147.83	153.39	+0.64
Fourteen	Boys.	934	"	151.28	0.197	6.03	4.0	101.8	6.19	4.3	146.00	151.62	157.36	+0.34
	Girls.	1088	"	152.58	0.160	5.29	3.5	101.2	5.39	3.7	148.30	153.56	158.54	−0.98
Fifteen	Boys.	495	"	158.43	0.321	7.15	4.5	102.3	7.15	4.7	151.81	158.79	166.04	+0.36
	Girls.	677	"	156.38	0.176	4.58	2.9	100.9	3.80	2.5	152.62	156.21	161.41	−0.17
Sixteen	Boys.	189	"	163.96	0.574	7.89	4.8	102.3	5.53	3.5	158.04	165.31	171.49	+1.35
	Girls.	413	"	158.51	0.217	4.41	2.8	100.7	2.13	1.4	155.39	159.31	163.26	+0.80
Seventeen	Boys.	75	"	168.56	0.581	5.03	3.0	102.1	4.60	2.8	164.77	169.58	175.38	+1.02
	Girls.	202	"	159.01	0.285	4.05	2.5	102.1	0.50	3.2	155.36	169.62	163.45	+0.61
Eighteen	Boys.	32	"	175.31	0.761	4.31	2.5	102.9	6.75	4.0	171.14	175.50	179.94	+0.19
	Girls.	164	"	160.47	0.334	4.28	2.7	101.3	1.46	0.9	166.65	161.38	165.41	+0.91
Nineteen	Girls.	83	"	158.45	0.517	4.71	2.9	101.8			155.17	158.90	163.09	+0.45
Twenty	Girls.	76	"	160.17	0.474	4.13	2.6	100.1			156.45	161.50	164.98	+1.33
Twenty-one	Girls.	37	"	161.27	0.712	4.33	2.7	100.1			157.40	160.88	166.68	−0.39

TABLE No. 42.
The Girth of Chest Midway Between Inspiration and Expiration.*

AGE AT NEAREST BIRTHDAY.	Sex.	Number of Observations.	Unit of Measurement.	Average.	Probable Error of Average. E.	Probable Deviation. d.	Relation of Probable Deviation to Avrage. d/A	Relation of Average to Height Standing.	Absolute Annual Increase of Average.	Relative Annual Increase of Average.	25 Percentile Grade.	Median or 50 Percentile Grade.	75 Percentile Grade.	Median Minus Average.
							%	%		%				
Six	Boys.	677	Centimetre.	59.05	0.083	2.22	3.8	54.2			57.45	59.32	61.42	+0.27
	Girls.	741	"	58.34	0.091	2.48	4.3	54.2			56.81	58.37	60.98	+0.53
Seven	Boys.	1708	"	60.62	0.057	2.38	3.9	53.1	1.57	2.7	58.79	60.89	63.25	+0.63
	Girls.	1631	"	59.47	0.062	2.47	4.2	52.7	1.13	1.9	57.65	59.87	62.31	+0.40
Eight	Boys.	2095	"	62.18	0.052	2.35	3.8	52.2	1.56	2.6	60.49	62.35	65.13	+0.37
	Girls.	2040	"	60.81	0.053	2.40	3.9	51.3	1.34	2.3	58.94	61.34	63.80	+0.53
Nine	Boys.	2120	"	63.90	0.055	2.51	3.9	51.4	1.72	2.8	61.97	64.33	66.91	+0.43
	Girls.	1966	"	62.51	0.057	2.53	4.0	50.6	1.70	2.8	60.65	63.03	65.60	+0.52
Ten	Boys.	1997	"	65.59	0.061	2.72	4.1	50.9	1.69	2.6	63.49	65.92	68.64	+0.33
	Girls.	1893	"	63.02	0.061	2.67	4.2	49.1	0.51	0.8	61.18	64.46	67.16	+0.45
Eleven	Boys.	1732	"	67.24	0.063	2.61	3.9	50.2	1.65	2.5	65.29	67.68	70.32	+0.44
	Girls.	1654	"	65.85	0.075	3.04	4.6	49.4	2.83	4.5	63.38	66.36	69.12	+0.52
Twelve	Boys.	1565	"	68.76	0.074	2.94	4.3	49.8	1.52	2.3	66.53	69.17	72.06	+0.41
	Girls.	1624	"	68.34	0.081	3.24	4.7	49.1	2.49	3.8	65.67	68.49	72.04	+0.16
Thirteen	Boys.	1228	"	70.61	0.089	3.11	4.4	49.4	1.85	2.7	68.09	71.01	74.13	+0.40
	Girls.	1313	"	71.29	0.098	3.54	5.0	48.6	2.95	4.3	68.32	71.64	75.13	+0.35
Fourteen	Boys.	925	"	73.27	0.118	3.58	4.9	49.3	2.66	3.8	70.24	73.38	77.27	+0.11
	Girls.	1020	"	74.13	0.114	3.65	4.9	49.5	2.84	4.0	71.03	74.59	78.29	+0.46
Fifteen	Boys.	498	"	76.56	0.169	3.77	4.9	49.4	3.29	4.5	73.25	76.94	80.82	+0.38
	Girls.	659	"	76.78	0.143	3.70	4.7	50.1	2.65	3.6	73.67	76.92	81.15	+0.15
Sixteen	Boys.	205	"	79.22	0.293	4.19	5.3	49.4	2.66	3.5	75.61	79.90	83.71	+0.68
	Girls.	397	"	78.85	0.164	3.27	4.1	50.5	2.07	2.7	75.98	79.22	82.85	+0.37
Seventeen	Boys.	90	"	81.39	0.353	3.15	3.9	49.3	2.17	2.7	78.90	82.20	85.77	+0.81
	Girls.	206	"	80.39	0.233	3.34	4.2	50.8	1.54	2.0	77.53	80.72	84.13	+0.34
Eighteen	Boys.	31	"	84.52	0.528	2.94	3.5	49.6	3.13	3.8	82.69	85.35	88.38	+0.83
	Girls.	162	"	80.45	0.254	3.23	4.0	49.7	0.06	0.1	77.68	80.92	84.90	+0.47
Nineteen	Girls.	82	"	79.10	0.380	3.43	4.3	50.3			76.61	79.81	82.84	+0.71
Twenty	Girls.	66	"	80.54	0.339	2.76	3.4				78.20	81.04	83.99	+0.61
Twenty-one	Girls.	32	"	81.51	0.521	2.82	3.5				78.63	81.49	84.99	—0.03

* Obtained by adding the Girth of Chest at full Inspiration to the Girth of Chest at full Expiration and dividing by 2.

Porter — The Growth of St. Louis Children. 355

TABLE No. 43.
THE STRENGTH OF SQUEEZE, RIGHT HAND.

AGE AT NEAREST BIRTHDAY.	Sex.	Number of Observations.	Unit of Measurement.	Average.	Probable Error of Average. E.	Probable Deviation. d.	Relation of Probable Deviation to Average. $\frac{d}{A}$	Relation of Average to Height Standing.	Absolute Annual Increase of Average.	Relative Annual Increase of Average.	Median or 50 Percentile Grade.	Median Minus Average.
					±	±	%	%		%		
Six	Boys.	626	Kilogram.	6.09	0.056	1.41	23.2	5.6			6.30	+0.21
	Girls.	687	"	5.14	0.053	1.39	27.0	4.8			5.93	+0.79
Seven	Boys.	1551	"	7.69	0.039	1.52	19.8	6.7	1.60	26.2	7.67	−0.02
	Girls.	1493	"	6.53	0.043	1.67	25.6	5.8	1.39	27.0	6.56	−0.03
Eight	Boys.	1880	"	9.38	0.046	1.95	20.8	7.9	1.69	22.0	9.59	+0.21
	Girls.	1873	"	8.11	0.043	1.87	23.1	6.9	1.58	24.2	8.24	+0.13
Nine	Boys.	2002	"	11.35	0.055	2.45	21.6	9.1	1.97	21.0	10.65	−0.70
	Girls.	1829	"	9.23	0.049	2.11	22.9	7.5	1.12	13.8	9.14	−0.09
Ten	Boys.	1878	"	12.83	0.061	2.66	20.7	10.0	1.48	13.0	12.72	−0.11
	Girls.	1801	"	10.42	0.053	2.27	21.9	8.1	1.19	12.9	10.36	−0.06
Eleven	Boys.	1644	"	14.37	0.068	2.74	19.1	10.7	1.54	12.0	14.29	−0.08
	Girls.	1613	"	11.80	0.060	2.42	20.5	8.9	1.36	13.1	12.07	+0.27
Twelve	Boys.	1506	"	16.70	0.082	3.17	18.8	12.1	2.33	16.2	16.56	−0.14
	Girls.	1553	"	13.46	0.072	2.83	21.0	9.7	1.66	14.1	13.39	−0.07
Thirteen	Boys.	1152	"	19.08	0.126	4.27	22.4	13.3	2.38	14.2	18.99	−0.09
	Girls.	1256	"	16.13	0.093	3.31	20.5	11.0	2.67	19.8	15.51	−0.62
Fourteen	Boys.	848	"	22.32	0.143	4.15	18.6	15.0	3.24	17.0	21.82	−0.50
	Girls.	950	"	18.02	0.133	4.11	22.8	12.0	1.89	11.7	17.78	−0.24
Fifteen	Boys.	447	"	26.69	0.244	5.10	19.1	17.2	4.37	16.6	26.17	−0.52
	Girls.	617	"	20.01	0.147	3.65	18.2	12.9	1.99	11.0	20.32	+0.31
Sixteen	Boys.	163	"	31.04	0.414	5.29	17.0	19.4	4.35	16.3	29.90	−1.14
	Girls.	356	"	21.78	0.200	2.77	17.3	13.8	1.77	8.8	21.60	−0.18
Seventeen	Boys.	21	"								30.70	
	Girls.	112	"	21.86	0.458			18.7	0.10	4.6	22.25	+0.37
Eighteen	Girls.	75	"	24.13				15.2	2.25	10.0	24.88	+0.75
Nineteen	Girls.	53	"					13.7			22.25	+0.43
Twenty	Girls.	61	"	21.82								
Twenty-one	Girls.	38	"	22.92				14.3			22.81	−0.11
											34.00	

356 *Trans. Acad. Sci. of St. Louis.*

TABLE No. 44.
THE STRENGTH OF SQUEEZE, LEFT HAND.

AGE AT NEAREST BIRTHDAY.	Sex.	Number of Observations.	Unit of Measurement.	Average.	Probable Error of Average. E.	Probable Deviation. d.	Relation of Probable Deviation to Av'age. $\frac{d}{A}$	Relation of Average to Height Standing.	Absolute Annual Increase Average.	Relative Annual Increase of Average.	Median or 50 Percentile Grade.	Median Minus Average.
					±	±	%	%		%		
Six	Boys.	629	Kilogram.	5.59	0.057	1.43	25.6	5.1			6.12	+0.53
	Girls.	686	"	4.77	0.056	1.47	30.8	4.4			5.46	+0.69
Seven	Boys.	1550	"	7.15	0.044	1.72	24.1	6.3	1.56	27.9	7.11	−0.04
	Girls.	1488	"	5.70	0.042	1.62	28.4	5.1	0.97	20.3	6.24	+0.54
Eight	Boys.	1882	"	8.76	0.048	2.08	23.2	7.4	1.61	22.5	8.74	−0.02
	Girls.	1882	"	7.52	0.046	2.00	26.6	6.4	1.82	31.9	7.30	−0.22
Nine	Boys.	2007	"	10.43	0.059	2.66	25.5	8.4	1.67	19.1	10.24	−0.19
	Girls.	1928	"	8.47	0.049	2.10	24.8	6.9	0.95	12.6	8.29	−0.18
Ten	Boys.	1886	"	11.72	0.059	2.57	21.9	9.1	1.29	12.4	11.59	−0.13
	Girls.	1798	"	9.38	0.053	2.23	23.8	7.3	0.91	10.7	9.97	−0.59
Eleven	Boys.	1650	"	13.49	0.067	2.70	20.0	10.1	1.77	15.1	13.26	−0.23
	Girls.	1616	"	11.07	0.059	2.36	21.3	8.3	1.69	18.0	11.09	+0.02
Twelve	Boys.	1502	"	15.37	0.080	3.11	20.2	11.1	1.88	13.9	15.08	−0.29
	Girls.	1554	"	12.60	0.076	2.98	23.7	9.1	1.53	13.8	12.65	+0.05
Thirteen	Boys.	1148	"	17.31	0.103	3.50	20.2	12.1	1.94	12.6	16.95	−0.36
	Girls.	1257	"	14.83	0.094	3.35	22.6	10.1	2.23	17.7	14.71	−0.12
Fourteen	Boys.	854	"	20.27	0.134	3.93	19.4	13.6	2.96	17.5	19.77	−0.50
	Girls.	947	"	17.13	0.127	3.90	22.7	11.4	2.30	15.5	16.44	−0.69
Fifteen	Boys.	439	"	23.94	0.243	5.10	21.3	15.4	3.67	18.1	24.04	+0.08
	Girls.	618	"	18.23	0.154	3.84	21.1	11.8	1.10	6.4	18.64	+0.41
Sixteen	Boys.	162	"	27.25	0.438	5.58	20.5	17.0	3.31	13.8	27.79	+0.54
	Girls.	359	"	19.86	0.207	3.93	19.8	12.6	1.63	8.9	19.78	−0.08
Seventeen	Boys.	58	"	29.58	0.639	4.87	16.5	17.9	2.33	12.8	30.18	+0.60
	Girls.	144	"	20.10	0.264	3.17	15.8	12.6	0.24	12.1	20.00	−0.10
Eighteen	Girls.	75	"	23.48				14.8	3.38	16.8	23.50	+0.02
Nineteen	Girls.	53	"	20.79				13.1			21.17	+0.38
Twenty	Girls.	62	"	22.19				13.9			22.67	+0.48
Twenty-one	Girls.	38	"	23.28							23.00	−0.28

TABLE No. 45.
THE STRENGTH OF SQUEEZE, RIGHT HAND.

AGE AT NEAREST BIRTHDAY.	Sex.	Unit of Measurement.	Dynamometers A. & D.				Dynamometer B.				Dynamometers C. & E.			
			No. of Observations.	Average.	Median.	Probable Deviation	No. of Observations.	Average.	Median.	Probable Deviation	No. of Observations.	Average.	Median.	Probable Deviation
Six........	Boys.	Kilo-gram.	159	5.25	5.74	1.24	198	10.25	10.61	2.04	269	16.03	15.82	4.08
	Girls.	"	197	4.25	5.32	1.29	201	8.68	9.53	2.21	289	13.91	15.80	3.60
Seven......	Boys.	"	436	7.26	7.24	1.86	427	12.29	11.85	1.32	688	19.67	20.24	4.52
	Girls.	"	377	6.35	6.49	1.85	395	10.18	10.62	2.11	721	16.64	15.88	4.47
Eight......	Boys.	"	744	9.85	10.21	2.47	377	14.15	14.70	2.84	759	22.87	22.72	4.44
	Girls.	"	746	8.15	8.57	2.41	393	12.22	12.30	2.57	734	20.67	20.53	3.85
Nine........	Boys.	"	877	12.29	11.25	3.10	351	16.47	15.96	3.33	774	27.77	26.00	5.25
	Girls.	"	829	9.39	9.28	2.70	328	13.33	13.33	2.92	672	24.17	23.78	5.62
Ten........	Boys.	"	941	13.65	12.89	3.23	363	18.94	20.02	3.57	574	31.50	30.59	6.11
	Girls.	"	847	10.81	10.64	2.52	395	15.21	15.57	3.13	559	26.48	25.65	5.59
Eleven.....	Boys.	"	823	14.64	15.20	3.05	340	21.99	20.96	4.04	481	35.63	35.26	6.40
	Girls.	"	761	11.92	11.72	2.46	373	17.53	18.62	3.82	479	30.36	30.38	5.78
Twelve.....	Boys.	"	678	17.24	16.98	3.36	347	25.90	25.75	4.93	481	40.35	40.16	7.43
	Girls.	"	730	13.71	13.81	3.01	313	20.46	20.04	3.95	510	33.60	32.61	7.27
Thirteen....	Boys.	"	534	18.82	19.11	3.75	257	30.36	30.47	5.60	361	46.22	45.42	9.14
	Girls.	"	575	16.52	15.96	3.41	261	24.97	25.15	4.72	420	39.38	35.96	8.55
Fourteen....	Boys.	"	399	22.61	22.28	4.12	189	36.26	35.52	6.97	260	51.83	50.74	9.65
	Girls.	"	491	18.63	18.55	4.20	171	28.27	28.92	5.92	288	42.98	40.45	10.60
Fifteen.....	Boys.	"	249	27.23	26.79	5.39	79	41.68	40.88	7.46	119	64.85	63.26	1.27
	Girls.	"	343	19.88	20.37	3.30	87	32.45	32.50	5.78	187	48.11	49.10	9.90
Sixteen.....	Boys.	"	102	30.42	30.55	5.29	31	43.32	41.85	13.59	30	86.40	80.00	1.37
	Girls.	"	219	20.81	20.90	3.30	43	37.00	36.50	6.17	94	51.48	51.00	10.08
Seventeen...	Boys.	"	21	37.38	30.70	3.78								
	Girls.	"	112	21.88	22.25									
Eighteen....	Girls.	"	75	24.13	24.88	4.82								
Nineteen....	Girls.	"	53	21.82	22.25	3.27								
Twenty.....	Girls.	"	61	22.92	22.81	2.19								
Twenty-one.	Girls.	"	38		24.00									

358   Trans. Acad. Sci. of St. Louis.

TABLE No. 46.

THE STRENGTH OF SQUEEZE, LEFT HAND.

AGE AT NEAREST BIRTHDAY.	Sex.	Unit of Measurement.	Dynamometers A. & D.				Dynamometer B.				Dynamometers C. & E.			
			No. of Observations.	Average.	Median	Probable Deviation.	No. of Observations.	Average.	Median	Probable Deviation.	No. of Observations.	Average.	Median	Probable Deviation.
Six	Boys.	Kilogram.	158	4.80	5.57	1.43	198	9.39	10.28	2.11	273	14.77	15.38	3.72
	Girls.	"	196	3.77	5.00	1.49	202	8.12	8.72	2.05	288	13.24	14.37	3.85
Seven	Boys.	"	440	6.75	7.02	1.72	423	11.26	10.96	2.38	687	18.56	18.09	4.74
	Girls.	"	376	5.53	5.87	1.57	393	8.42	10.35	2.56	719	15.30	15.47	4.06
Eight	Boys.	"	741	9.27	9.39	2.64	379	13.10	13.10	2.54	762	21.35	20.96	4.77
	Girls.	"	745	7.90	7.49	2.43	404	10.88	10.77	2.26	733	19.08	18.62	5.18
Nine	Boys.	"	876	11.14	10.68	3.06	361	15.11	15.22	3.09	780	25.94	25.48	7.15
	Girls.	"	830	8.44	8.54	2.43	328	12.30	12.27	2.72	670	22.20	20.97	5.14
Ten	Boys.	"	947	12.31	11.62	3.14	365	17.26	17.21	3.65	574	29.25	30.03	5.58
	Girls.	"	847	10.06	10.30	2.64	392	14.15	14.76	3.08	559	24.82	25.14	5.04
Eleven	Boys.	"	824	14.04	13.66	3.20	338	19.94	20.25	3.97	488	33.87	32.60	5.83
	Girls.	"	762	10.95	10.81	2.54	374	16.52	16.57	3.62	480	28.90	29.29	5.41
Twelve	Boys.	"	673	15.86	15.61	3.20	348	23.65	23.71	5.24	481	37.47	35.84	6.86
	Girls.	"	732	12.71	12.60	3.30	310	18.80	20.22	4.16	512	32.34	30.78	7.24
Thirteen	Boys.	"	535	17.47	17.32	3.69	268	27.39	26.83	5.52	355	41.98	40.58	8.05
	Girls.	"	574	15.25	15.24	3.48	261	23.05	23.05	5.26	422	35.82	35.08	7.80
Fourteen	Boys.	"	402	20.46	20.68	3.90	191	32.66	31.44	5.96	261	48.27	45.90	10.01
	Girls.	"	490	17.13	17.06	3.92	171	26.31	26.00	4.98	286	43.18	38.75	11.20
Fifteen	Boys.	"	247	24.39	24.69	4.74	69	40.12	39.50	8.42	125	54.03	55.00	12.74
	Girls.	"	343	18.05	18.25	3.45	87	28.89	30.32	6.26	188	44.99	45.29	10.00
Sixteen	Boys.	"	191	28.17	29.50	4.91	20	40.33	39.25	10.17	31	68.52	70.20	13.15
	Girls.	"	221	19.34	19.29	3.24	42	31.74	32.75	5.85	96	49.41	47.40	12.66
Seventeen	Boys.	"	36	30.97	31.67	4.87					22	68.77	70.00	11.79
	Girls.	"	112	19.79	19.40	3.24					32	49.78	50.25	7.54
Eighteen	Girls.	"	75	23.48	23.50	5.05								
Nineteen	Girls.	"	53	20.79	21.17	3.40								
Twenty	Girls.	"	62	22.19	22.67	2.23								
Twenty-one	Girls.	"	38	23.28	23.00	2.88								

TABLE No. 47.
THE LENGTH OF HEAD.

AGE AT NEAREST BIRTHDAY.	Sex.	Number of Observations	Unit of Measurement	Average.	Probable Error of Average. E.	Probable Deviation. ±d	Relation of Probable Deviation to Av'rage. d/A %	Absolute Average Relation of Height to Standing. %	Average. Annual Increase of.	Relative Annual Increase of Average. %	25 Percentile Grade.	Median or 50 Percentile Grade.	75 Percentile Grade.	Median Minus Average.
Six	Boys.	606	Milli-metre.	178.39	0.171	4.21	2.36	16.4			174.63	178.86	182.79	+0.47
	Girls.	606	"	173.45	0.170	4.20	2.42	16.1			170.19	174.34	178.14	+0.89
Seven	Boys.	1493	"	178.54	0.118	4.57	2.56	15.7	0.15	0.08	174.84	179.92	183.79	+0.38
	Girls.	1511	"	174.09	0.108	4.22	2.42	15.4	0.64	0.37	171.41	175.58	179.76	+0.49
Eight	Boys.	2079	"	179.62	0.106	4.85	2.70	15.1	1.08	0.60	176.03	180.44	184.87	+0.82
	Girls.	2125	"	175.18	0.092	4.26	2.43	14.8	1.09	0.63	172.17	176.41	180.82	+1.23
Nine	Boys.	1986	"	180.72	0.109	4.87	2.69	14.5	1.10	0.61	176.71	181.00	185.69	+0.28
	Girls.	1884	"	176.39	0.101	4.39	2.49	14.3	1.21	0.69	173.69	177.80	181.68	+1.41
Ten	Boys.	1912	"	181.45	0.094	4.13	2.28	14.1	0.73	0.40	178.10	181.51	185.85	−0.06
	Girls.	1790	"	177.24	0.102	4.33	2.44	13.8	0.85	0.48	174.15	178.52	182.47	+1.28
Eleven	Boys.	1654	"	182.37	0.129	5.26	2.89	13.6	0.92	0.50	178.85	182.86	187.26	+0.49
	Girls.	1560	"	178.08	0.119	4.70	2.64	13.4	0.84	0.47	174.71	179.75	183.78	+1.67
Twelve	Boys.	1576	"	182.84	0.115	4.56	2.49	13.2	0.47	0.26	179.19	183.08	187.52	+0.24
	Girls.	1516	"	179.50	0.114	4.46	2.49	12.9	1.42	0.80	176.43	180.57	184.86	+1.07
Thirteen	Boys.	1207	"	183.84	0.131	4.55	2.48	12.9	1.00	0.55	179.93	184.21	188.78	+0.37
	Girls.	1187	"	181.44	0.132	4.55	2.51	12.4	1.94	1.08	178.35	182.48	187.05	+1.04
Fourteen	Boys.	890	"	186.93	0.182	5.44	2.91	12.6	3.09	1.68	182.08	186.24	190.75	−0.69
	Girls.	1008	"	183.41	0.143	4.53	2.47	12.2	1.97	1.09	180.07	184.71	185.78	+1.30
Fifteen	Boys.	502	"	187.01	0.208	4.66	2.49	12.1	0.08	0.04	183.01	187.69	191.97	+0.68
	Girls.	649	"	185.12	0.180	4.58	2.47	11.9	1.71	0.93	181.53	185.67	190.90	+0.55
Sixteen	Boys.	191	"	189.06	0.355	4.93	2.61	11.8	2.05	1.10	184.79	190.04	194.77	+0.98
	Girls.	400	"	186.84	0.256	5.12	2.74	12.1	1.72	0.93	183.02	187.70	192.47	+0.86
Seventeen	Boys.	78	"	189.45	0.575	5.08	2.68	11.6	0.39	0.21		190.93		+1.48
	Girls.	221	"	188.14	0.275	4.09	2.17	11.8	1.30	0.70		188.60		+0.46
Eighteen	Boys.	32	"	193.91	0.217	4.13	2.13	11.4	4.46	2.35		193.75		−0.16
	Girls.	161	"	187.97	0.252	3.20	1.70	11.9				188.41		+0.44
Nineteen	Girls.	77	"	187.91	0.466	4.09	2.18	11.8			184.36	188.31	191.75	+0.40
Twenty	Girls.	75	"	187.81	0.433	3.75	2.01	11.7			184.72	188.08		+0.27
Twenty-one	Girls.		"											

TABLE No. 48.
THE WIDTH OF HEAD.

AGE AT NEAREST BIRTHDAY	Sex	Number of Observations	Unit of Measurement	Average	Probable Error of Average E	Probable Deviation d	Relation of Probable Deviation to Average d/A %	Average Relation of Height to Standing	Absolute Annual Increase of Average	Relative Annual Increase of Average %	25 Percentile Grade	Median or 50 Percentile Grade	75 Percentile Grade	Median Minus Average	Width Length Index
Six	Boys	573	Millimetre	143.29	0.118	2.82	1.97	13.2			140.77	143.76	147.11	+0.47	0.80
	Girls	609		140.27	0.154	3.81	2.72	13.0			137.21	140.83	144.54	+0.56	0.81
Seven	Boys	1571	"	144.37	0.090	2.58	2.48	12.7	1.08	0.75	141.57	144.98	148.26	+0.61	0.81
	Girls	1505		141.40	0.099	3.86	2.73	12.5	1.13	0.81	138.63	142.10	145.44	+0.70	0.81
Eight	Boys	1992	"	145.30	0.082	2.68	2.53	12.3	0.93	0.64	142.37	145.63	149.43	+0.33	0.81
	Girls	1985		142.31	0.082	3.66	2.57	12.0	0.91	0.64	139.50	142.84	146.30	+0.53	0.81
Nine	Boys	1962	"	145.87	0.084	3.74	2.56	11.7	0.57	0.39	142.79	146.08	150.05	+0.21	0.81
	Girls	1914		143.04	0.083	3.65	2.55	11.6	0.73	0.51	140.47	143.58	147.01	+0.54	0.81
Ten	Boys	1827	"	146.59	0.090	3.85	2.63	11.4	0.72	0.49	143.40	147.11	151.05	+0.52	0.81
	Girls	1803		143.76	0.093	3.96	2.76	11.2	0.71	0.50	140.99	144.64	148.02	+0.79	0.81
Eleven	Boys	1561	"	147.29	0.095	3.76	2.55	11.0	0.70	0.48	144.30	147.50	151.42	+0.31	0.81
	Girls	1541		145.05	0.097	3.80	2.62	10.9	1.31	0.90	142.06	145.54	149.33	+0.49	0.81
Twelve	Boys	1531	"	147.98	0.092	3.61	2.44	10.7	0.69	0.47	145.19	148.36	152.08	+0.38	0.81
	Girls	1460		145.64	0.105	4.03	2.77	10.5	0.59	0.42	142.91	146.31	150.06	+0.67	0.81
Thirteen	Boys	1175	"	148.73	0.114	3.90	2.62	10.4	0.75	0.51	145.46	149.01	152.98	+0.27	0.81
	Girls	1167		146.78	0.115	3.92	2.67	10.0	1.14	0.78	143.44	147.15	151.17	+0.37	0.81
Fourteen	Boys	873	"	149.50	0.129	3.82	2.56	10.1	0.77	0.52	146.04	150.04	153.42	+0.54	0.80
	Girls	927		147.90	0.130	3.96	2.68	9.8	1.12	0.76	144.57	148.25	152.14	+0.35	0.81
Fifteen	Boys	469	"	149.63	0.186	4.03	2.69	9.7	1.13	0.87	146.24	150.41	154.04	+0.78	0.80
	Girls	658		148.29	0.150	3.84	2.59	9.6	0.39	0.26	145.16	148.71	152.63	+0.42	0.80
Sixteen	Boys	195	"	150.98	0.249	3.48	2.31	9.4	1.35	0.90	147.75	151.35	154.96	+0.37	0.80
	Girls	395		148.96	0.210	4.17	2.80	9.4	0.66	0.44	145.85	149.67	153.44	+0.72	0.80
Seventeen	Boys	75	"	152.09	0.486	4.21	2.77	9.2	1.11	0.74	149.84	152.50	156.55	+0.41	0.80
	Girls	221		150.04	0.242	3.60	2.40	9.4	1.09	0.73	146.99	150.43	154.38	+0.39	0.80
Eighteen	Boys	32	"	151.66	0.685	3.88	2.56	8.9				152.00		+0.34	0.78
	Girls	165		149.09	0.300	3.85	2.58	9.4				150.16	163.21	+1.07	0.79
Nineteen	Girls	79	"	149.11	0.446	4.14	2.78	9.4			145.96	150.17	154.10	+1.06	0.79
Twenty	Girls	73	"	147.82	0.426	3.64	2.46	9.2			144.69	147.70	152.26	−0.12	0.79
Twenty-one	Girls		"												

TABLE No. 49.

The Height of Face from Root of Nose to Point of Chin.

Age at Nearest Birthday	Sex	Number of Observations	Unit of Measurement	Average	Probable Error of Average E.	Probable Deviation d.	Relation of Probable Deviation to Average d/A %	Relation of Average to Height Standing %	Absolute Annual Increase of Average	Relative Annual Increase of Average %	25 Percentile Grade	Median or 50 Percentile Grade	75 Percentile Grade	Median Minus Average
Six	Boys	604	Millimetre	93.72	0.207	5.09	5.23	8.6			89.24	93.20	99.45	−0.52
	Girls	612		91.42	0.194	4.80	5.25	8.5			86.88	90.89	96.46	−0.53
Seven	Boys	1580	"	95.87	0.120	4.80	5.06	8.4	2.15	2.89	91.28	95.57	101.11	−0.30
	Girls	1509		93.77	0.115	4.45	4.75	8.3	2.35	2.57	89.35	93.20	98.38	−0.57
Eight	Boys	2057	"	97.98	0.103	4.67	4.77	8.2	2.11	2.20	93.73	97.78	102.71	−0.20
	Girls	2016		95.65	0.099	4.46	4.66	8.1	1.88	2.01	91.67	95.60	99.79	−0.05
Nine	Boys	2011	"	99.51	0.107	4.82	4.84	8.0	1.53	1.56	95.04	99.42	104.25	−0.09
	Girls	1898		97.85	0.103	4.48	5.76	7.9	2.20	2.30	93.61	97.60	102.31	−0.25
Ten	Boys	1868	"	101.06	0.107	4.77	4.72	7.8	1.55	1.56	96.66	100.73	105.93	−0.33
	Girls	1820		99.39	0.109	4.64	4.67	7.8	1.54	1.57	95.04	99.02	104.22	−0.37
Eleven	Boys	1650	"	103.87	0.127	5.18	5.01	7.5	2.31	2.29	98.63	103.23	108.85	−0.14
	Girls	1572		101.44	0.116	4.61	4.54	7.6	2.05	2.06	97.08	101.20	106.32	−0.24
Twelve	Boys	1568	"	104.25	0.132	5.22	5.01	7.4	0.88	0.85	98.63	104.45	109.77	+0.20
	Girls	1623		103.46	0.129	5.05	4.88	7.4	2.02	1.99	98.81	103.53	108.83	+0.07
Thirteen	Boys	1205	"	106.24	0.160	5.54	5.21	7.3	1.99	1.91	101.64	106.34	111.76	+0.10
	Girls	1231		105.92	0.133	4.68	4.42	7.2	2.46	2.38	101.60	105.71	111.15	−0.21
Fourteen	Boys	893	"	108.43	0.174	6.21	4.80	7.2	2.19	2.06	103.64	108.51	113.79	+0.08
	Girls	997		107.87	0.156	4.94	4.58	7.2	1.95	1.84	103.37	107.72	112.86	−0.15
Fifteen	Boys	479	"	111.63	0.238	5.21	4.67	7.1	3.20	2.95	106.04	111.52	116.88	−0.11
	Girls	672		109.79	0.192	4.98	4.53	7.1	1.92	1.78	104.56	109.40	114.05	−0.39
Sixteen	Boys	191	"	113.63	0.346	4.79	4.22	7.0	2.00	1.79	109.19	113.73	118.98	+0.10
	Girls	403		110.48	0.245	4.91	4.45	6.9	0.69	0.63	105.92	109.83	115.48	−0.61
Seventeen	Boys	78	"	117.56	0.744	6.57	5.59	7.1	3.93	3.46	112.42	118.50	122.89	+0.94
	Girls	223		110.04	0.345	5.16	4.68	6.9			105.11	110.50	115.46	+0.46
Eighteen	Boys	31	"	120.84	0.883	4.92	4.16	7.1	3.28	2.79		121.00		+0.16
	Girls	163		109.77	0.297	3.79	3.45				105.65	110.04		+0.27
Nineteen	Girls	80	"	108.72	0.421	3.77	3.50	6.8			104.18	108.34	112.90	−0.38
Twenty	Girls	73	"	110.26	0.469	4.01	3.64	6.9			106.34	110.30	114.13	+0.04
Twenty-one	Girls	34	"	110.06	0.627	3.66	3.32					110.00		−0.06

The Width of Face.

AGE AT NEAREST BIRTHDAY.	Sex.	Number of Observations.	Unit of Measurement.	Average.	Probable Error of Average. E.	Probable Deviation. d.	Relation of Probable Deviation to Av'rge. d/A %	Relation of Average to Height Standing. %	Absolute Annual Increase of Average.	Relative Annual Increase of Average. %	25 Percentile Grade.	Median or 50 Percentile Grade.	75 Percentile Grade.	Median Minus Average.	Width Height Index.	Width Height (Hair-Line) Index.
Six	Boys.	608	Millimetre.	117.24	0.172	±4.24	3.62	10.8			114.04	118.20	121.62	+0.96	1.25	0.77
	Girls.	608		115.21	0.174	4.30	3.73	10.7			111.92	115.89	119.39	+0.68	1.26	0.77
Seven	Boys.	1637	"	117.78	0.101	4.09	3.47	10.3	0.54	0.46	114.41	118.40	121.97	+0.62	1.23	0.76
	Girls.	1516		116.11	0.101	3.92	3.38	10.3	0.90	0.78	112.95	116.38	120.22	+0.27	1.24	0.76
Eight	Boys.	2062	"	118.82	0.090	4.08	3.44	10.0	1.04	0.88	115.65	119.75	122.96	+0.93	1.21	0.75
	Girls.	2065		117.63	0.087	3.97	3.38	9.9	1.52	1.31	113.62	117.45	120.99	−0.18	1.23	0.76
Nine	Boys.	1927	"	119.91	0.101	4.44	3.70	9.6	1.09	0.92	116.54	120.75	124.59	+0.84	1.21	0.75
	Girls.	1905		118.02	0.092	4.02	3.41	9.5	0.39	0.33	114.59	118.81	122.18	+0.79	1.21	0.75
Ten	Boys.	1922	"	121.60	0.092	4.05	3.33	9.4	1.69	1.41	118.02	121.63	125.32	+0.03	1.20	0.75
	Girls.	1829		119.49	0.127	4.38	3.67	9.3	1.47	1.25	116.19	120.42	124.04	+0.93	1.20	0.75
Eleven	Boys.	1666	"	122.73	0.108	4.52	3.68	9.2	1.13	0.93	119.44	123.22	126.89	+0.49	1.19	0.75
	Girls.	1504		121.24	0.107	4.17	3.44	9.1	1.75	1.46	117.82	121.71	125.33	+0.47	1.20	0.75
Twelve	Boys.	1525	"	123.81	0.113	4.42	3.57	9.0	1.08	0.88	120.90	124.59	128.45	+0.78	1.19	0.75
	Girls.	1526		122.44	0.109	4.26	3.48	8.8	1.20	0.99	119.50	123.07	127.01	+0.63	1.18	0.74
Thirteen	Boys.	1913	"	125.83	0.122	4.27	3.39	8.8	2.02	1.63	121.96	126.02	129.79	+0.19	1.19	0.75
	Girls.	1248		124.15	0.127	4.45	3.59	8.5	1.71	1.40	120.62	124.86	128.82	+0.71	1.17	0.74
Fourteen	Boys.	898	"	126.81	0.153	4.56	3.59	8.5	0.98	0.78	123.17	127.32	131.64	+0.51	1.17	0.74
	Girls.	997		125.67	0.135	4.26	3.39	8.3	1.52	1.22	122.31	126.51	130.49	+0.84	1.16	0.72
Fifteen	Boys.	485	"	128.32	0.196	4.33	3.37	8.3	1.51	1.19	124.02	129.03	133.72	+0.71	1.12	0.74
	Girls.	678		127.85	0.157	4.08	3.19	8.2	2.18	1.73	124.51	128.30	132.02	+0.45	1.16	0.72
Sixteen	Boys.	193	"	130.27	0.360	5.01	3.85	8.2	1.95	1.52	126.39	130.82	134.89	+0.55	1.15	0.73
	Girls.	409		129.47	0.201	4.07	3.14	8.2	1.62	1.26	126.21	130.18	133.60	+0.71	1.17	0.73
Seventeen	Boys.	79	"	131.96	0.527	4.69	3.56	8.0	1.69	1.30		133.75	135.00	+1.79	1.12	0.72
	Girls.	219		130.63	0.275	4.08	3.12	8.2	1.16	0.90	127.91	131.38		+0.75	1.19	0.73
Eighteen	Boys.	33	"	135.00	0.718	4.13	3.06		3.04	2.01		135.75	135.00	+0.75	1.12	0.74
	Girls.	163		131.42	0.265	3.38	2.57	8.2	0.79	0.60	129.14	131.97		+0.55	1.20	0.73
Nineteen	Girls.	80	"	130.24	0.420	3.76	2.89	8.3				130.44		+0.20	1.20	0.72
Twenty	Girls.	73	"	130.75	0.419	3.58	2.74	8.2				132.04		+1.29	1.19	0.73
Twenty-one	Girls.	34	"		0.528	3.08		8.2				132.00				

TABLE No. 51.
THE HEIGHT OF FACE FROM HAIR-LINE TO POINT OF CHIN.

AGE AT NEAREST BIRTHDAY.	Sex.	Number of Observations.	Unit of Measurement.	Average.	Probable Error of Average. E.	Probable Deviation. d.	Relation of Probable Deviation to Average. $\frac{d}{A}$	Relation of Average to Height Standing.	Absolute Annual Increase of Average.	Relative Annual Increase of Average.	25 Percentile Grade.	Median or 50 Percentile Grade.	75 Percentile Grade.	Median Minus Average.
							%	%		%				
Six........	Boys.	611	Millimetre.	152.68	0.209	5.18	3.4	14.0			148.64	153.02	158.50	+0.34
	Girls.	609	"	150.16	0.238	5.88	3.9	14.0			144.68	150.45	156.21	+0.39
Seven.....	Boys.	1621	"	154.57	0.159	6.40	4.1	13.6	1.89	1.40	150.13	155.33	160.40	+0.76
	Girls.	1486	"	153.36	0.137	5.28	3.4	13.6	3.20	2.13	148.73	154.19	159.32	+0.83
Eight......	Boys.	2012	"	157.50	0.122	5.46	3.5	13.2	2.93	1.36	153.04	158.04	163.10	+0.54
	Girls.	1965	"	155.19	0.119	5.27	3.4	13.1	1.83	1.19	151.05	155.60	161.11	+0.41
Nine.......	Boys.	1997	"	159.43	0.127	5.67	3.6	12.8	1.93	1.32	154.45	160.25	165.62	+0.82
	Girls.	1989	"	157.44	0.121	5.38	3.4	12.8	2.25	1.45	152.31	157.76	163.40	+0.32
Ten........	Boys.	1909	"	161.37	0.127	5.55	3.4	12.5	1.94	1.28	156.47	161.70	167.38	+0.33
	Girls.	1835	"	160.04	0.131	5.62	3.5	12.5	2.60	1.65	155.36	160.56	166.15	+0.52
Eleven.....	Boys.	1653	"	163.39	0.143	5.80	3.5	12.2	2.02	1.25	158.67	164.11	169.45	+0.72
	Girls.	1681	"	162.55	0.143	5.70	3.5	12.2	2.51	1.57	157.38	163.40	169.53	+0.85
Twelve.....	Boys.	1577	"	165.23	0.137	5.44	3.3	11.9	1.84	1.22	160.37	165.62	171.54	+0.39
	Girls.	1510	"	165.29	0.148	5.76	3.5	11.9	2.74	1.69	160.27	165.95	171.40	+0.66
Thirteen...	Boys.	1211	"	167.62	0.173	6.01	3.6	11.7	2.39	1.20	162.59	168.16	173.87	+0.54
	Girls.	1220	"	168.35	0.173	6.04	3.6	11.5	3.06	1.85	163.18	169.15	174.85	+0.80
Fourteen...	Boys.	896	"	170.49	0.236	7.05	4.1	11.5	2.87	1.17	165.01	171.04	177.42	+0.55
	Girls.	998	"	171.38	0.193	6.11	3.6	11.4	3.08	1.80	163.66	172.00	178.28	+0.62
Fifteen....	Boys.	482	"	174.20	0.292	6.41	3.7	11.3	3.81	1.15	168.74	175.08	181.23	+0.78
	Girls.	656	"	174.20	0.211	5.40	3.1	11.2	2.82	1.65	168.92	174.85	180.10	+0.65
Sixteen....	Boys.	193	"	178.19	0.425	5.90	3.3	11.1	3.89	1.13	172.76	177.93	184.22	−0.26
	Girls.	395	"	176.28	0.286	5.69	3.2	11.2	2.08	1.19	171.32	176.81	183.09	+0.53
Seventeen.	Boys.	79	"	182.28	0.353	7.60	4.2	11.0	4.09	1.11		182.25		−0.03
	Girls.	201	"	178.51	0.437	6.20	3.5	11.2	2.23	1.26	173.62	179.41	184.90	+0.90
Eighteen..	Boys.	33	"	183.67	0.139	8.00	4.4	10.8	1.39	1.10		185.66		+1.99
	Girls.	139	"	180.97	0.520	6.13	3.4	11.4	2.46	1.38	176.32	181.93	186.81	+0.96
Nineteen...	Girls.	69	"	181.91	0.749	6.22	3.4	11.4	0.94	0.52	174.29	180.08	185.70	−1.83
Twenty....	Girls.	72	"	179.09	0.560	4.76	2.7	11.2			174.80	179.00	184.38	−0.09
Twenty-one	Girls.	33	"	179.97	0.786	4.52	2.5					181.75		+1.78

BIBLIOGRAPHY.

ANGERSTEIN (W.): Die Maasverhältnisse des menschlichen Körpers und das Wachsthum der Knaben. Köln, 1865.

ANUTSCHIN: On the geographical distribution of the male population of Russia in respect of height. Transactions of the Imperial Geographical Society, Vol. VII, Part 1 (Russian, cited by Sack).

BAXTER (J. H.): Medical statistics of the Provost-Marshal-General's Bureau, 1875, Vol. I.

BELYAIEFF: Materials for investigating the influence of schools on the physical development of pupils. Inaugural Dissertation, St. Petersburg, 1888 (Russian, cited by Sack).

BERTILLON (A.): Les proportions du corps humain. Revue Scientifique, Paris, 1889, t. 43, No. 17, pp. 524-529.

BEYER (H. G.): Observations on normal growth and development of the human body under systematised exercise. Report of the chief of the Bureau of Medicine and Surgery to the Secretary of the Navy. Washington, 1893, pp. 141-160.

BLAGOVIDOFF (I.): Materials for the investigation of the health of the Mongolian-Asiatic races (inorodze) in the province of Simbirsk. Inaugural Dissertation, St. Petersburg, 1886 (Russian, cited by Sack).

BOAS (F.): The growth of children. Science, Vol. XIX, No. 483, pp. 256-257, May 6, 1892, and No. 485, pp. 281-282, May 20.

BOAS (F.): The growth of children. Science, Dec. 23, 1892. p. 351.

BOULTON: Some anthropometrical observations. British Medical Journal, March 4, 1876. pp. 280-282.

BOWDITCH (H. P.): The growth of children. Eighth Annual Report of the State Board of Health of Massachusetts. Boston, 1877.

BOWDITCH (H. P.): The growth of children, a supplementary investigation, with suggestions in regard to methods of research. Tenth Report of Board of Health of Massachusetts. Boston, 1879. pp. 33-62.

BOWDITCH (H. P.): The physique of women in Massachusetts. Twenty-first Report of the Board of Health of Massachusetts, 1889. Boston, 1890. pp. 287-304.

BOWDITCH (H. P.): The growth of children studied by Galton's Method of Percentile Grades. Twenty-second Report of Board of Health of Massachusetts, 1889-90. Boston, 1891, pp. 479-522.

BRESLAU: Neue Ergebnisse aus Schädelmessung an Neugebornen. Wiener medicinische Wochenschrift, Dec. 13, 1862, No. 50. S. 785-787.

BRUNNICHE: Ein Beitrag zur Beurtheilung der Körperentwickelung der Kinder. Journal für Kinderkrankheiten, 1866.

COWELL: Parliamentary Reports, 1833, XX, D. 1. p. 87.

CNOPF: Die Anwendung der Wage in der Kinderpraxis. Journal für Kinderkrankheiten. Bd. LVIII, 1872, S. 219-234.

DAFFNER (F.): Ueber Grösse, Gewicht, Kopf- und Brustumfang beim männlichen Individuum vom 13. bis 22. Lebensjahre, nebst vergleichender Angabe einiger Kopfmaasse. Archiv für Anthropologie, Braunschweig, Bd. XV, 1885, Supplement. S. 121–126.

DEMENTIEFF (E. M.): The development of muscular force in connection with general physical development. Inaugural Dissertation, Moscow, 1889 (Russian, cited by Sack).

DIEK: Materials for the investigation of height, weight etc., in childhood and youth. Inaugural Dissertation. St. Petersburg, 1883 (Russian, cited by Sack).

ERISMANN (F.): Untersuchungen über die körperliche Entwickelung der Arbeiterbevölkerung in Zentralrussland. Tübingen, 1889. Reprint from Archiv für soziale Gesetzgebung and Statistik.

FERGUS (W.) AND RODWELL (G. F.): On a series of measurements for statistical purposes, recently made at Marlborough College. The Journal of the Anthropological Institute of Great Britain and Ireland, London, May 12, 1874. pp. 126–130.

GALTON (F.): On the height and weight of boys, aged 14 years, in town and country public schools. The Journal of the Anthropological Institute of Great Britain and Ireland, Vol. V, 1876. pp. 174–180.

GALTON (F.): Natural Inheritance. London, 1889.

GARSON ET AL.: Report of the committee appointed for the purpose of calculating the anthropological measurements taken at the Newcastle meeting of the Association in 1889. Report of the British Association for the Advancement of Science, 1890, London, 1891, LX, pp. 549–552.

GAUSS: Bestimmung der Genauigkeit der Beobachtungen. Zeitschrift für Astronomie, 1816, März u. April, S. 185–197.

GEISSLER (A.) UND UHLITZSCH (R.): Die Grössenverhältnisse der Schulkinder im Schulinspectionsbezirk Freiburg. Separat-Abdruck aus Heft I und II d. Jahrgang XXXIV, 1888, der Zeitschrift des königlichen sächsischen statistischen Bureaus. S. 28–40.

GEISSLER (A.): Messungen von Schulkindern in Gohlis-Leipzig. Zeitschrift für Schulgesundheitspflege. Hamburg u. Leipzig, 1892, V, S. 249–253.

GRATSIANOFF: Materials for the investigation of physical development in childhood and youth in relation to heredity and to progress in school work. From observations made in Arzamas, province of Novgorod. Inaugural Dissertation, St. Petersburg, 1889.

GREENWOOD (J. M.): Heights and weights of children. Twentieth Annual Report of the Board of Education of the Kansas City Public Schools, 1890–91. pp. 48–56.

GREENWOOD (J. M.): Heights and weights of children. American Public Health Association Report, 1891, Concord, 1892, XVII. pp. 199–204.

GRINEVSKI (A.): On the physical development of children. Odessa, 1892.

HANSEN (S.): Ueber die individuellen Variationen der Körperproportionen. Archiv für Anthropologie, Bd. XX, 1892, S. 321–323.

HERSCHEL (SIR J. F. W.): Review of Lettres à S. A. R. le Duc règnant de Saxe-Cobourg et Gotha sur la Théorie des Probabilités appliquée aux Sciences Morales et Politiques par M. A. Quetelet, Astron. Royal de la Belgique. Edinburg Review, No. 185. July, 1850. pp. 1–57.

HITCHCOCK (E.): Comparative study of measurements of male and female students at Amherst, Mt. Holyoke and Wellesley Colleges. Physique, London, 1891, I, pp. 90–94.

v. IHERING: Zur Einführung von Oscillations-Exponenten in die Cranometrie. Archiv für Anthropometrie, Bd. X, S. 411–413.

JANSEN (A.): Étude d'anthropométric médicale, au point de vue de l'aptitude au service militaire. Mém. couron. de l'Academie royale de médicine de Belgique. Bruxelles, 1882.

KEY (A.): Die Pubertätsentwickelung. Verhandlungen des X. Internationalen medicinischen Congresses, Berlin, 1890, Bd. 1, S. 67.

KEY (A.): Schulhygienische Untersuchungen. Edited by L. Burgerstein. Hamburg and Leipzig, 1889. Original: Läroverkskomiténs betänkande III. Bilaga E. Kongl. boktryckeriet. Stockholm, 1885.

KIRCHHOFF (A.): Zur Statistik der Körpergrösse in Halle, dem Saalkreise und dem Mansfelder Seekreise. Archiv für Anthropologie, Braunschweig, 1892-3, XXI, S. 133–143.

KOTELMANN (L.): Die Körperverhältnisse der Gelehrtenschüler des Johanneums in Hamburg. Berlin, 1879. Separat-Abdruck aus der Zeitschrift des königlichen preussichen statistischen Bureaus, Jahrgang 1879.

LANDSBERGER: Das Wachsthum im Alter der Schulpflicht. Archiv für Anthropologie, Braunschweig, Bd. XVII, 1887, S. 229–264.

LESHAFT: Materials for the study of the years of school life. Health, 1879-1880, No. 127-131 (Russian, cited by Sack).

LEXIS (W.): Ueber die Theorie der Stabilität statistischer Reihen. Hildebrandt's Jahrbücher für Nationalökonomie, Bd. XXXII, S. 60–97, Jena, 1879.

LIHARZIK (F.): Das Gesetz des menschlichen Wachsthums. Wien, 1858.

LIHARZIK (F.): Der Bau und das Wachsthum der Menschen. Sitzungs-Berichte d. königlichen Akademie in Wien, math.-naturw. Klasse, Abtheilung II, Bd. 44, 1861.

LIHARZIK (F.): Das Gesetz des Wachsthumes und der Bau der Menschen, die Proportionslehre aller menschlichen Körpertheile für jedes Alter und für beide Geschlechter. Imp.-Folio. Wien, 1862.

MALLING-HANSEN (R.): Ueber Periodicität im Gewicht der Kinder an täglichen Wägungen wahrgenommen. Fragment I, Kopenhagen, 1883.

MALLING-HANSEN (R.): Perioden im Gewicht der Kinder und in der Sonnenwärme. Fragment III A, und Fragment III B, Kopenhagen, 1886.

MASHKOFFSKY: The measurement of the chest in healthy individuals and in consumptives. Inaugural Dissertation. St. Petersburg, 1889 (Russian, cited by Sack).

MERESHOFFSKY (K.): On the results and methods of the investigation of the physical development of children. Brochure (Russian, cited by Sack).

MIKHAILOFF: Materials for the estimation of physical development and disease in the village schools of the district of Russkis in the province of Moscow. Moscow, 1887 (Russian, cited by Sack).

MINOT (C. S.): Human growth. Boston Medical and Surgical Journal, July 22, 1881.

NAGORSKY: The influence of schools on the physical development of children. St. Petersburg, 1881 (Russian, cited by Sack).

PAGLIANI (L.): Sopra alcuni fattori dello sviluppo umano. Richerche Anthropometriche. Atti della r. Accad. di Scienze di Torino, Vol. XI, 1875-1876, pp. 694-760, e Arch. di Antrop. ed Etnol. italiana, Vol. VI, 1876, p. 129-183.

PAGLIANI (L.): Die Entwickelung des Menschen in den der Geschlechts - reife vorangehenden späteren Kinderjahren und im Jünglingsalter (von 7 bis 20 Jahren).

PAGLIANI (L.): Lo sviluppo umano per età, sesso, condizione sociale ed etnica studiato nel peso, statura, circonferenza toracica, capacità vitale e forza muscolare. Giornale della Società italiana d'igiene, Milano, 1879, I, No. 4, pp. 357-376; No. 5, pp. 453-491, No. 6, pp. 589-608.

PECKHAM (G. W.): The growth of children. Sixth Annual Report of the State Board of Health of Wisconsin, 1881. Madison, 1882, pp. 28-73.

PESKOFF (P. A.): Report of the commission for the inspection of factories in Moscow. Moscow, 1881. (Russian.)

PORTER (W. TOWNSEND): The physical basis of precocity and dullness. Transactions of the Academy of Science of St. Louis. Vol. VI, No. 7, 1893, pp. 161-181.

PORTER (W. TOWNSEND): Untersuchungen der Schulkinder in Bezug auf die physischen Grundlagen ihrer geistigen Entwickelung. Verhandlungen der Berliner anthropologischen Gesellschaft, 15 Juli, 1893. Zeitschrift für Ethnologie, S. 337-354.

PORTER (W. TOWNSEND): The relation between the growth of children and their deviation from the physical type of their sex and age. Transactions of the Academy of Science of St. Louis, Vol. VI, No. 10, pp. 233-250.

QUETELET (A.): Recherches sur le poids de l'homme aux différents âges. Bull. de l'Académie royale des sciences, des lettres et des beaux-arts de Belgique. Bruxelles, 1832-1834, I, pp. 20-21.

QUETELET (A.): Sur l'homme et le développement de ses facultés. Paris, 1835.

QUETELET (A.): Lettres à S. A. R. le Duc régnant de Saxe-Cobourg et Gotha, sur la théorie des probabilités appliquée aux sciences morales et politiques. Bruxelles, 1846.

QUETELET (A.): Sur les proportions des hommes, qui se font remarquer par un excès ou un défaut de taille. Bull. de l'Académie royale des sciences, dés lettres, et des beaux-arts de Belgique, 1847, t. IV. pp. 138-142.

QUETELET (A.): De la statistique considérée sous le rapport du physique, du moral et de l'intelligence de l'homme. Bulletin de la commission centrale de statistique, Bruxelles, 1860, tome VIII, pp. 433-467.

QUETELET (A.): Physique sociale. Bruxelles, 1869.

QUETELET (A.): Anthropométrie. Bruxelles, 1870.

ROBERTS (C.): A manual of anthropometry. London, 1878.

RUDIN (Wd. W.): Ueber einen Versuch der Anwendung der Anthropometrie zur Beurtheilung der physischen Ausbildung der Zöglinge einer Turnschule in Mologa. VIII. Congress russischer Naturforscher und Aerzte in St. Petersburg, 1890. Archiv für Anthropologie, 1891, p. 382.

RUMA (R.): Anthropometrical materials for the determination of the physical development of pupils. Collection of works on legal medicine. 1880. Vol. III (Russian), pt. 2, pp. 95-131.

RUSSOW (A.): Vergleichende Beobachtungen über den Einfluss der Ernährung mit der Brust und der künstlichen Ernährung auf das Gewicht und den Wuchs (Länge) der Kinder. Jahrbuch für Kinderheilkunde, Leipzig, 1880-1, n. F. XVI, 86-132.

SACK: Physical development of the children in the Middle Schools of Moscow, 1892. (Russian.)

SHBANKOFF: The influence of the Common School (Volksschule) on the physical development of pupils. The Messenger for Legal Medicine, 1889, III (Russian, cited by Sack).

SHEBOLDAEFF: Sanitary questions in the Common Schools (Volksschulen) of the Konotopski Semstvo Tshernigoff, 1887 (Russian, cited by Sack).

SOAMES (H. A.): The scientific measurement of children. London, 1891.

SOGRAF (N. J.): Anthropometrical researches in the provinces Jaroslav, Kostroma and Vladimir. VIII Congress of Russian Naturalists and Physicians in St. Petersburg, 1890.

STEET (G. C.) Notes on the development and growth of boys between thirteen and twenty years of age. St. George's Hospital Reports, London, 1874-6.

STEVENSON (Wm.): On the relation of weight to height and the rate of growth in man. The Lancet, London, Sept. 22, 1888, pp. 560-564.

STIEDA (L.): Ueber die Anwendung der Wahrscheinlichkeitsrechnung in der anthropologischen Statistik. Archiv für Anthropologie. Braunschweig, Bd. XIV, 1882-3, S. 167-182.

SULIGOFFSKI (F.): Some remarks on anthropometrical measurements of young men in the male gymnasium in Radom. Medicine, Warsaw, 1887, XV, pp. 512, 528, 544, 559, 641.

THOMA (R.): Untersuchungen über die Grösse und das Gewicht der anatomischen Bestandtheile des menschlichen Körpers im gesunden und kranken Zustande. Leipzig, 1982.

TOPINARD (P.): Étude sur la taille, considerée suivant l'âge, le sexe, l'individu et les races. Paris, 1876.

v. VIERORDT (K.): Grundriss der Physiologie des Menschen. Tübingen, 1871.

v. VIERORDT (K.): Physiologie des Kindesalters, in Gerhardt's Handbuch der Kinderkrankheiten. Tübingen, 1877, Bd. I, S. 83.

VAHL (M.): Om Vejning af Börn. Reprint, Copenhagen, 1881.

VAHL (M.): Mitteilungen über das Gewicht nicht erwachsener Mädchen [1874-83]. Vortrag auf dem Aerzte Kongress zu Kopenhagen, 1884.

WEST (G. M): The growth of the breadth of the face. Science, New York, 1891, No. 18, pp. 10-11.

WEST (G. M.): Worcester school children — the growth of the body, head and face. Science, Vol. XXI, No. 518, Jan. 6, 1893, pp. 2-4.

WEST (G. M.): Anthropometrische Untersuchungen über die Schulkinder in Worcester, Mass. Archiv für Anthropologie. Braunschweig. July, 1893. S. 13-48.

WORONICHIN (N.): Fortlaufende Wägungen während der Dentition. Jahrbuch für Kinderheilkunde. Leipzig, 1880-1, n. F. XVI. S. 133-143.

WRETLIND: Jakttagelser rörande helsotillstaandet i naagra af Göteborgs flickskolor. Eira, 1878.

YASTSHINSKY: Anthropometrical materials for the study of the development of height, weight etc. in Poles and Jews during their school years (Russian, cited by Sack).

ZEISING: Neue Lehre von den Proportionen des menschlichen Körpers. Leipzig, 1854.

ZEISING: Verhandlungen der Kaiserl.-Leopold.-carolinischen Akademie der Naturforscher. Bd. 26, 1858.

ZUBKOFFSKY: An investigation of the sanitary condition of the Military Gymnasium in Polozk. Inaugural Dissertation. St. Petersburg, 1879 (Russian, cited by Sack).

CORRIGENDA.

Page 263. Author's name omitted: W. Townsend Porter.
   288. Boys, aged 16, hair-line to point of chin should read 0.425 mm.
   297. Boys, aged 10, median minus average should read — 0.07 cm.
   313. Boys, aged 8, 50 percentile grade should read 119.77 cm.
   313. Girls, aged 8, 50 percentile grade should read 118.74 cm.
   315. Girls, aged 6, 50 percentile grade should read 107.29 cm.
   315. Girls, aged 8, 50 percentile grade should read 118.33 cm.
   321. Girls, aged 14, number of observations should read 997.

LIST OF PLATES.

PLATE I. The calculated and the observed distribution of the height standing of 2192 St. Louis school girls, aged 8.
II. Median minus average values (weight, height, height sitting, span of arms).
III. Median minus average values (chest-girth and strength of squeeze).
IV. Median minus average values (head and face measurements).
V. Boys' weight (percentile grades).
VI. Girls' weight (percentile grades).
VII. Boys' height standing (percentile grades).
VIII. Girls' height standing (percentile grades).
IX. Boys' height sitting (percentile grades).
X. Girls' height sitting (percentile grades).
XI. Boys' span of arms (percentile grades).
XII. Girls' span of arms (percentile grades).
XIII. Boys' girth of chest (percentile grades).
XIV. Girls' girth of chest (percentile grades).
XV. Boys' length of head (percentile grades).
XVI. Girls' length of head (percentile grades).
XVII. Boys' width of head (percentile grades).
XVIII. Girls' width of head (percentile grades).
XIX. Boys' height of face from root of nose to point of chin (percentile grades).
XX. Girls' height of face from root of nose to point of chin (percentile grades).
XXI. Boys' width of face (percentile grades).
XXII. Girls' width of face (percentile grades).
XXIII. Boys' height of face from hair-line to point of chin (percentile grades).
XXIV. Girls' height of face from hair-line to point of chin (percentile grades).
XXV. Weight, 75, 50 and 25 percentile grades.
XXVI. Height standing, 75, 50 and 25 percentile grades.
XXVII. Height sitting, 75, 50 and 25 percentile grades.
XXVIII. Span of arms, 75, 50 and 25 percentile grades.
XXIX. Girth of chest, 75, 50 and 25 percentile grades.
XXX. Mean expansion of chest.
XXXI. Mean strength of squeeze.
XXXII. Length of head, 75, 50 and 25 percentile grades.
XXXIII. Width of head, 75, 50 and 25 percentile grades.
XXXIV. Height of face from root of nose to point of chin, 75, 50 and 25 percentile grades.

PLATE XXXV. Width of face, 75, 50 and 25 percentile grades.
XXXVI. Height of face from hair-line to point of chin, 75, 50 and 25 percentile grades.
XXXVII. Percentile grades, weight.
XXXVIII. Percentile grades, height standing.
XXXIX. Percentile grades, height sitting.
XL. Percentile grades, span of arms.
XLI. Percentile grades, girth of chest.
XLII. Absolute annual increase (height standing, weight, span of arms).
XLIII. Absolute annual increase (height sitting, girth of chest, strength of squeeze).
XLIV. Relative annual increase (strength of squeeze, weight, height standing, height sitting, span of arms, chest-girth).
XLV. Relation of average weight, span of arms, girth of chest, etc., to average height.
XLVI. The weights of daughters of manual tradesmen distributed by school grade.

INDEX OF TABLES.

No. 1. The distribution of 2000 measurements of the same quantity when the deviation of the individual observations from the true value of the measured quantity is due to purely accidental causes... 277
2. Heights of United States recruits................................. 278
3. Observed distribution of the heights of 2192 St. Louis school girls, aged 8.. 279
4. The calculation of the average height of St. Louis school girls, aged 9... 280
5. The calculation of the probable deviation (d) from the average height (118.36 cm.) of 2193 girls, aged 8........................ 283
6. Stieda's table for calculating the number of observations at any distance from the mean or average within the limits: $M + 5d$ and $M - 5d$... 284
7. The theoretical and the observed distribution of the heights of 2192 girls, aged 8.. 286
8. The percentile distribution of the heights of girls, aged 9...... 287
9. The probable error of the average................................. 288
10. The probable deviation (d) from the average.................. 291
11. Median minus average values...............................296, 297
12. Sums of median minus average values........................... 300
13. A comparison of weights of the daughters of manual tradesmen with the weights of the daughters of professional men and merchants.. 305
14. The percentile distribution by school grade of the daughters of merchants and professional men (i. e. favored classes) compared with that of the daughters of manual tradesmen (handworkers)... 307
15. The weights of girls whose parents were born in Germany compared with the weights of girls whose parents were born in the United States.. 309
16. The weights of boys whose parents were born in Germany compared with the weights of boys whose parents were born in the United States.. 310
17. The weight (percentile grades).................................... 312
18. The height standing (percentile grades).......................... 313
19. The height sitting (percentile grades)............................ 314
20. The span of arms (percentile grades)............................. 315
21. The girth of chest at full inspiration (percentile grades)...... 316
22. The girth of chest at full expiration (percentile grades)...... 317
23. The girth of chest midway between full inspiration and full expiration (percentile grades)... 318
24. The length of head (percentile grades)........................... 319

(373)

	PAGE.
No. 25. The width of head (percentile grades)	320
26. The height of face from root of nose to point of chin (percentile grades)	321
27. The width of face (percentile grades)	322
28. The height of face from hair-line to point of chin (percentile grades)	323
29. Ages at which girls begin and cease to be larger than boys	324
30. The absolute annual increase in height standing	328
31. The absolute annual increase in weight	329
32. The absolute annual increase in span of arms	330
33. The absolute annual increase in height sitting	331
34. The absolute annual increase in girth of chest	332
35. Median weight of boys aged 11 distributed by school grade	335
36. Median weight of the daughters of manual tradesmen distributed by school grade	337
37. Median weight of the daughters of professional men distributed by school grade	338
38. The height standing	350
39. The weight	351
40. The height sitting	352
41. The span of arms	353
42. The girth of chest midway between inspiration and expiration	354
43. The strength of squeeze, right hand	355
44. The strength of squeeze, left hand	356
45. The strength of squeeze, right hand	357
46. The strength of squeeze, left hand	358
47. The length of head	359
48. The width of head	360
49. The height of face from root of nose to point of chin	361
50. The width of face	362
51. The height of face from hair-line to point of chin	363

GENERAL INDEX.

	PAGE.
Accidental influences	275
Acuteness of vision*	271
American girls' weights	309
American boys' weights	310

Annual increase—
 Absolute...........................327, 328 to 332, Plates XLII, XLIII
 Relative..327, 333, Plate XLIV
Anthropometrical series... 277
Anthropometrical system in schools.............................341, 347
Apparatus.. 269
Assistants—
 Names.. 264
 Duties...268, 269
Authorization by School Board... 263
Average—
 Calculation.. 280
 Definition... 280
 Deviation from.. 291
 Error of..287, 288
 Of different series.. 281
 Relation to mean.. 294

Bibliography.. 364
Birth-place of pupils... 309
Bowditch on median and average.. 294

Chart of days... 268
Chest expansion.. Plate XXX
Chest-girth at full expiration—
 Median minus average..............................296, 300, Plate III
 Method of measuring..271, 274
 Percentile grades..311, 317
Chest-girth at full inspiration —
 Median minus average..............................296, 300, Plate III
 Method of measuring..271, 274
 Percentile grades..311, 316

---

* The acuteness of vision of St. Louis school children will be discussed in a separate paper.

General Index.

PAGE.

Chest-girth midway between full inspiration and full expiration —
   Absolute annual increase.................................332, Plate XLIII
   Median minus average...296, 354
   Percentile grades..........311, 318, 354, Plates XIII, XIV, XXIX, XLI
   Probable deviation...291, 354
   Probable error..288, 354
   Relation to height standing........................334, 354, Plate XLV
   Relation of probable deviation to average............................ 354
   Relative annual increase..... ....327, 333, 354, Plate XLIV
Collection of data... 265
Constant causes... 275
Curves all printed.. 264

Deafness unsuspected.. 273
Difference of individual from type..... 292
Distribution of observations —
   About middle value... 277
   According to Thoma.. 277
   Girls' heights... 279
   Of 1000 individuals. 341
   Theoretical and observed...................... ....283, 286, Plate I
Duration of investigation.. 268

Errors —
   Constant..275, 292
   Of observation... 292
   Physiological........................... 292
   Variable..275, 292

Face (see Height of face and Width of face).

Galton's percentile grades..286, 287
Generalizing method..263, 294, 339
German parentage and weight.....................................309, 310
Girls larger than boys of same age...............................324, 325

Head (see Length of head and Width of head).
Head measurers 269
Hearing tests 272
Hearing tests impracticable... 273
Height—
   Abnormal height a disadvantage.............................333, 343
   Basis of a system of standards................................... 342
   Of United States recruits.. 278
Height of face from hair-line to point of chin —
   Absolute annual increase... 363
   Median minus average....................................296, Plate IV
   Method of measuring... 278
   Percentile grades.........311, 323, 363, Plates XXIII, XXIV, XXXVI

General Index. 377

	PAGE.

Height of face from hair-line to point of chin — *Continued.*
  Probable deviation...291, 363
  Probable error..288, 363
  Relation of probable deviation to average.........................363
  Relation to height standing.......................334, 363, Plate XLV
  Relative annual increase..363
Height of face from root of nose to point of chin —
  Absolute annual increase..361
  Median minus average......................................296, Plate IV
  Method of measuring...273
  Percentile grades................311, 321, 361, Plates XIX, XX, XXXIV
  Probable deviation...291, 361
  Probable error...288, 361
  Relation to probable deviation to average.........................361
  Relation to height standing......................334, 361, Plate XLV
  Relative annual increase..361
Height sitting —
  Absolute annual increase...........................327, 331, Plate XLIII
  Median minus average......................................296, 352, Plate II
  Method of measuring...270
  Percentile grades..........311, 314, 352, Plates IX, X, XXVII, XXXIX
  Probable deviation...291, 352
  Probable error...288, 352
  Relation of probable deviation to average.........................352
  Relation to height standing......................334, 352, Plate XLV
  Relative annual increase....................327, 333, 352, Plate XLIV
Height standing —
  Absolute annual increase..........................327, 328, Plate XLII
  Distribution of, in girls aged 8...........................286, Plate I
  Median minus average......................................296, 350, Plate II
  Method of measuring...270
  Percentile grades.....311, 313, 350, Plates VII, VIII, XXVI, XXXVIII
  Probable deviation...291, 350
  Probable error...288, 350
  Ratio to other physical measurements.....................334, Plate XLV
  Relation of probable deviation to average.........................350
  Relative annual increase....................327, 333, 350, Plate XLIV
Herschel on means...294

Indices —
  Cranial...360
  Facial..362
Individualizing method.....................................294, 339, 340
Influences affecting measurements...............................275, 276
Instructions to observers...269

Length of head —
  Absolute annual increase..359
  Median minus average......................................296, Plate IV
  Method of measuring...272

Length of head — *Continued.*
  Percentile grades..............311, 319, 359, Plates XV, XVI, XXXII
  Probable deviation..291, 359
  Probable error..288, 359
  Relation of probable deviation to average........................ 352
  Relation to height standing.....................334, 359, Plate XLV
  Relative annual increase... 359
Loans and gifts.. 264

Material of investigation —
  Limitations.. 294
  Trustworthiness.. 290
Manual tradesmen... 303
Manual tradesmen's daughters —
  Relative number..307, 308
  Weights..305, 337
Mean (see Median value).
Median value —
  Advantage over average.......................................294, 298
  Application to individuals..........................293, 294, 339, 340
  Calculation.. 281
  Definition...279, 281
  Minus average....................296, 298, 299, 300, Plates II, III, IV
  Uses limited... 294
Mental labor and physical development.............................. 335
Method of collecting measurements................................. 263
Middle value —
  Definition... 279
  Uses... 293

Nationality, influence on weight................................... 309
Number of children measured.. 268

Occupation of parents.. 302
Overstrain... 346

Percentile grades—
  Calculation.. 286
  Plates...V to XLI, inclusive
  Tables...312 to 323 and 350 to 363, inclusive
  Uses... 311
Plates, List of.. 371
Printed "forms" employed......................................265, 267, 269
Probable deviation—
  Calculation.. 282
  Definition... 282
  From average... 291
  From mean.. 284
  Table.. 291
Probable error of average—
  Calculation.. 287
  Table.. 288

	PAGE.
Professional men and merchants	302
Daughters' weights	305, 306, 338

Quetelet's law.... 278

Rate of growth. 327

Schools measured. 266
Sexual differences in growth. 324
Social status influences weight. 305, 310, 336, 338
Span of arms —
   Absolute annual increase. 327, 330, Plate XLII
   Median minus average. 296, Plate II
   Method of measuring. 270
   Percentile grades. 311, 315, 353, Plates XI, XII, XXVIII, XL
   Probable deviation. 291, 353
   Probable error. 288, 353
   Relation of probable deviation to average. 353
   Relation to height standing. 334, 353
   Relative annual increase. 327, 333, 353, Plate XLIV
Squeeze (see Strength of squeeze).
Statistical methods employed. 275, 279, 289
Strength of squeeze —
   Absolute annual increase. 327, 355, 356, Plate XLIII
   Dynamometer tables. 357, 358
   Median minus average. 296, 355, 356, Plate III
   Method of measuring. 271
   Percentile grades. 355, 356, Plate XXXI
   Probable deviation. 291, 355, 356
   Probable error. 288, 355, 356
   Relation of probable deviation to average. 355, 356
   Relation to height standing. 334, 355, 356
   Relative annual increase. 327, 333, 355, 356, Plate XLIV
Successful pupils larger than unsuccessful. 338

Tables, Index of. 373
Types. 277, 289, 293, 301

Weight —
   Absolute annual increase. 327, 329, Plate XLII
   Median minus average. 296, 351, Plate II
   Method of weighing. 271
   Percentile grades. 311, 312, 351, Plates V, VI, XXV, XXXVII
   Probable deviation. 291, 351
   Probable error. 288, 351
   Relation of probable deviation to average. 351
   Relation to height standing. 334, 343, 351, Plate XLV
   Relative annual increase. 327, 333, 351, Plate XLIV

Width of face —
Weight of American girls 309
    German girls 309
    Manual tradesmen's daughters .... 338, Plate XLVI
    Successful and unsuccessful pupils 336, 337
    Absolute annual increase 362
    Indices of width — height 362
    Median minus average .... 296, 362, Plate IV
    Method of measuring 273
    Percentile grades .... 311, 322, 362, Plates XXI, XXII, XXXV
    Probable deviation 291, 362
    Probable error 289, 362
    Relation of probable deviation to average 362
    Relation to height standing .... 311, 334, 362, Plate XLV
    Relative annual increase 362
Width of head —
    Absolute annual increase 360
    Index of width — length 360
    Median minus average .... 296, 360, Plate IV
    Method of measuring 273
    Percentile grades .... 311, 320, 360, Plates XVII, XVIII, XXXIII
    Probable deviation 291, 360
    Probable error 288, 360
    Relation of probable deviation to average 360
    Relation to height standing .... 334, 360, Plate XLV
    Relative annual increase 360
Working-plan 265

Issued April 14, 1894.

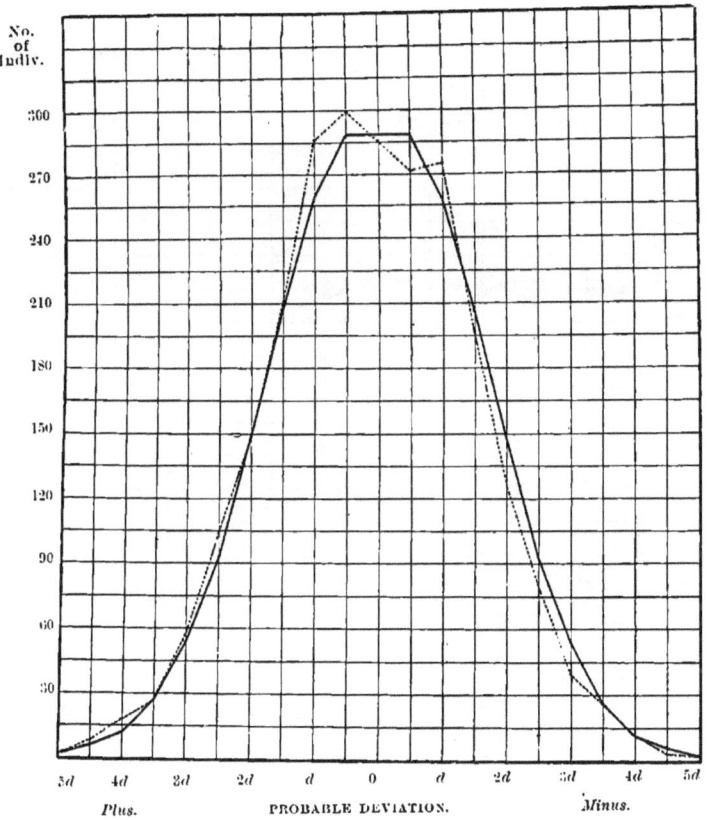

Vol. VI. No. 12. Plate I (from Table No. 7; page 286).

The calculated and the observed distribution of the height standing of 2192 St. Louis School Girls, aged 8.

Unbroken Line:
Distribution according to theory.

Broken Line:
Distribution according to observation.

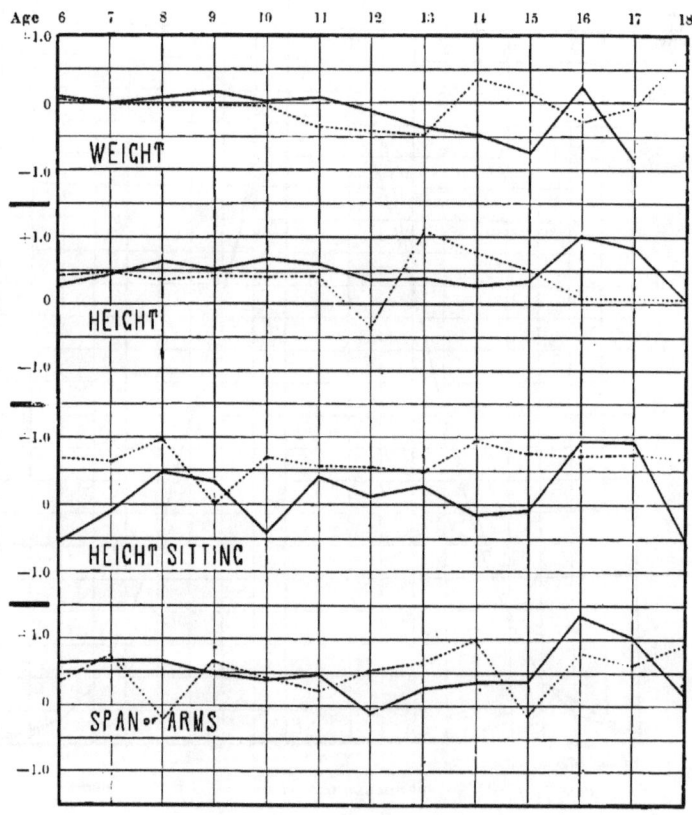

VOL. VI. NO. 12. PLATE II (from Table No. 11; pages 296, 299).
Median Minus Average Values.
Unbroken Line: Boys.   Broken Line: Girls.

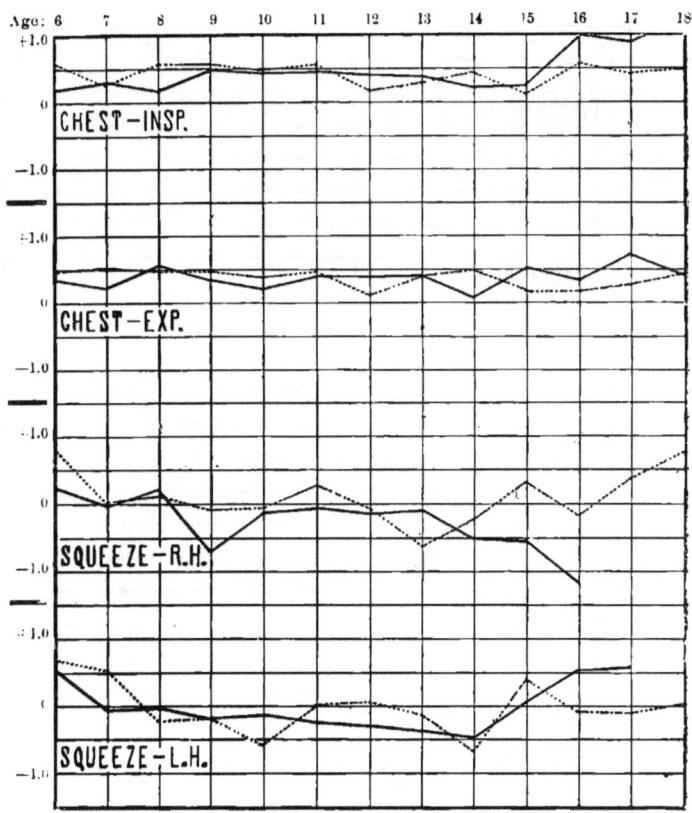

VOL. VI. NO. 12. PLATE III (from Table No. 11; pages 296, 299).
Median Minus Average Values.
Unbroken Line: Boys.   Broken Line: Girls.

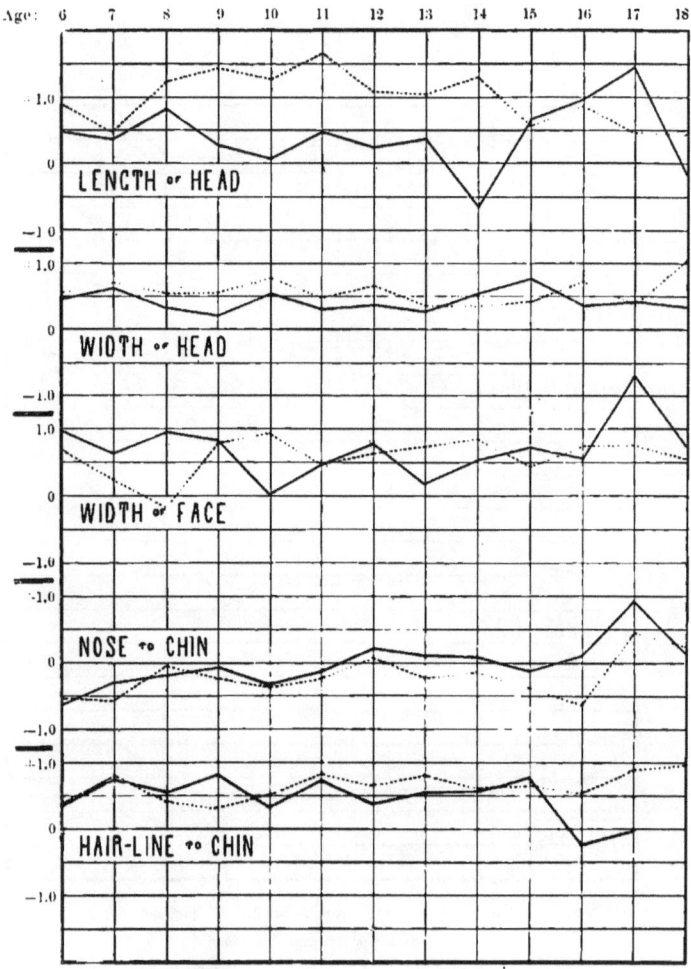

VOL. VI. NO. 12. PLATE IV (from Table No. 11; pages 296, 299).
Median Minus Average Values.
Unbroken Line: Boys.   Broken Line: Girls.

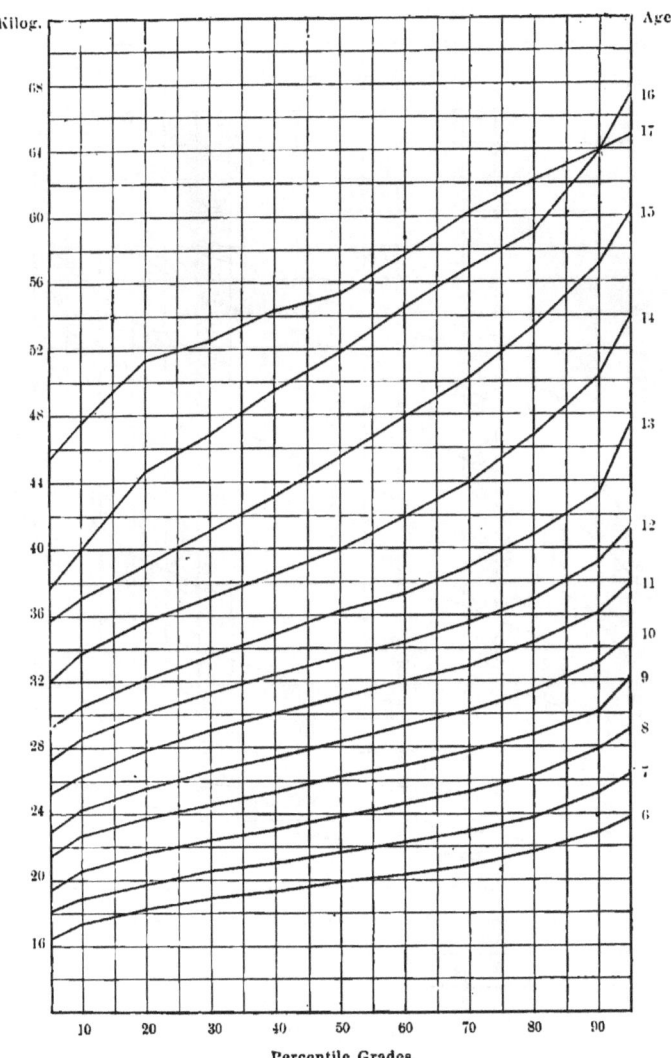

VOL. VI. NO. 12. PLATE V (from Table No. 17; pages 311, 312).
Boys' Weight.

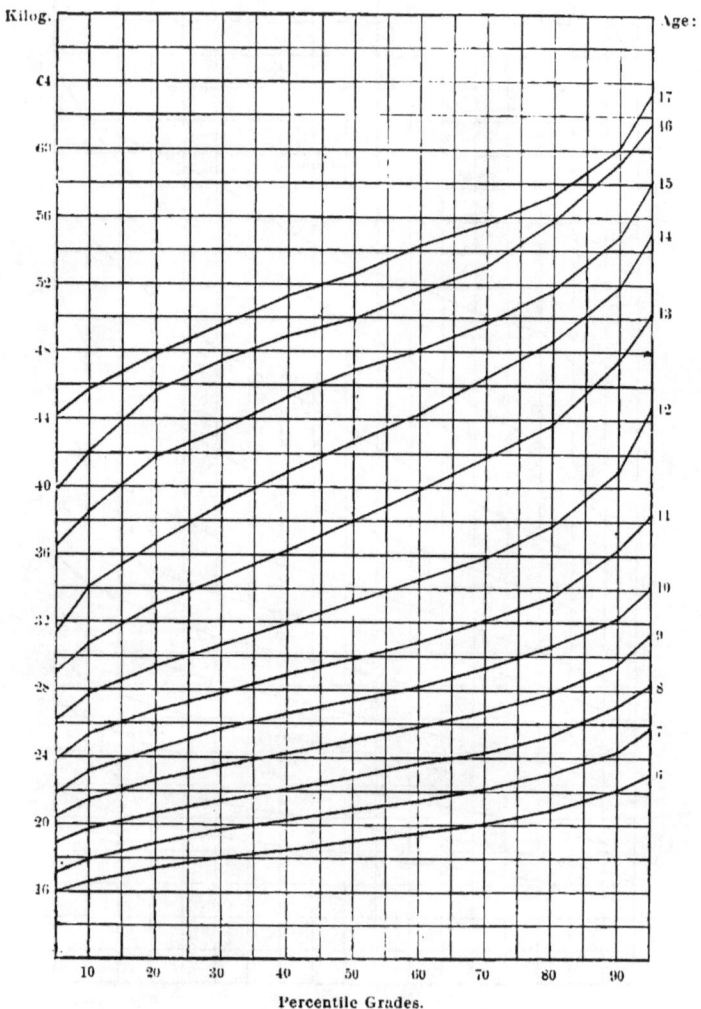

PLATE VI (from Table No. 17; pages 311, 312).
Girls' Weight.

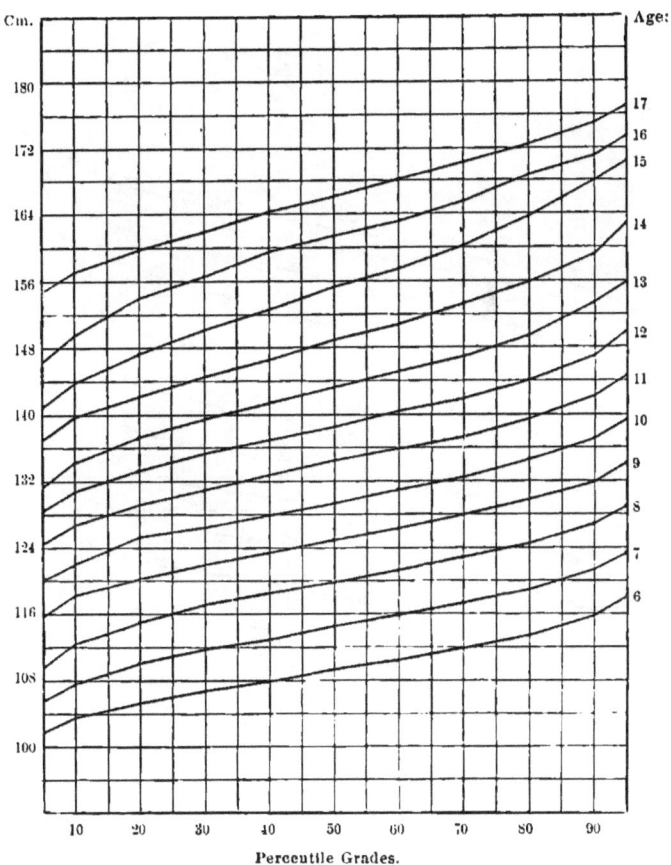

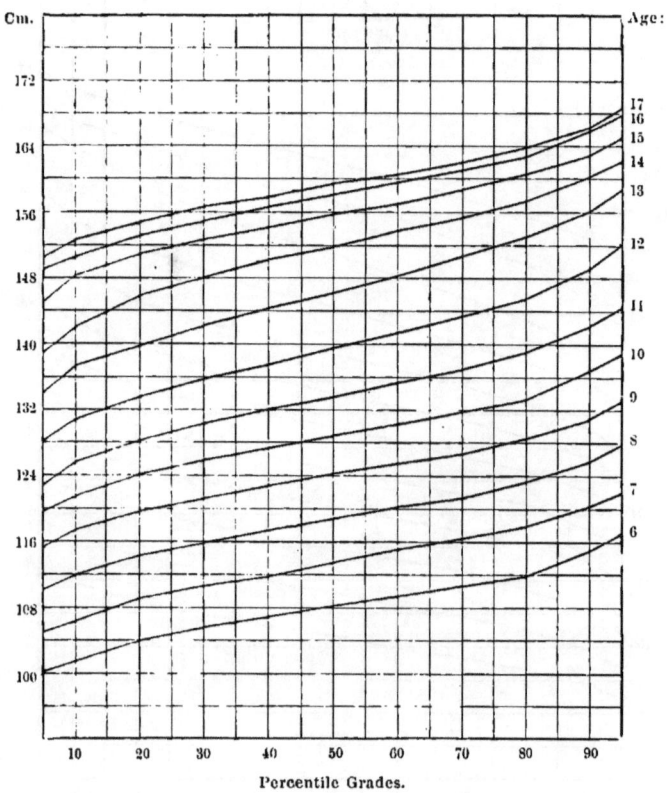

VOL. VI. NO. 12. PLATE VIII (from Table No. 18; pages 311, 313).
Girls' Height Standing.

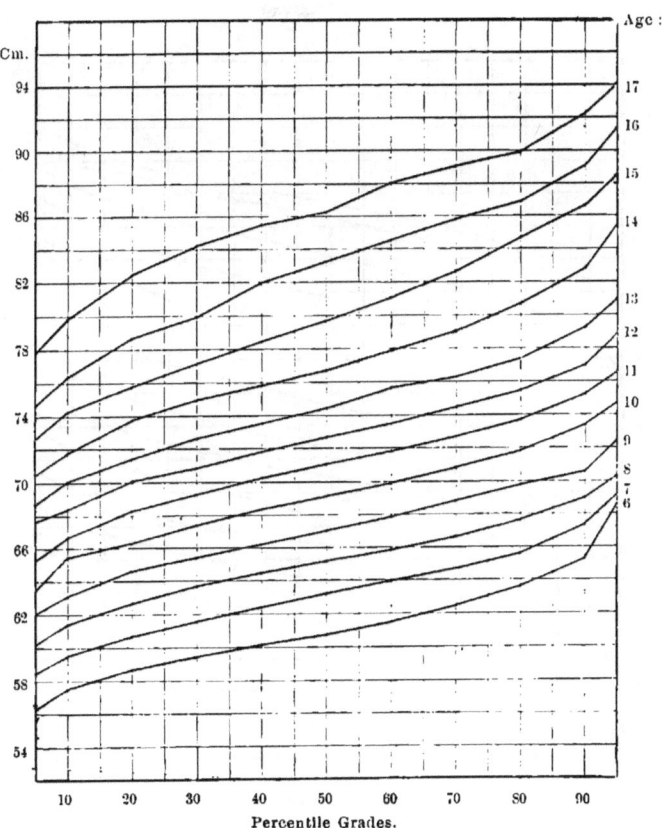

VOL. VI. NO. 12. PLATE IX (from Table No. 19; pages 311, 314).
Boys' Height Sitting.

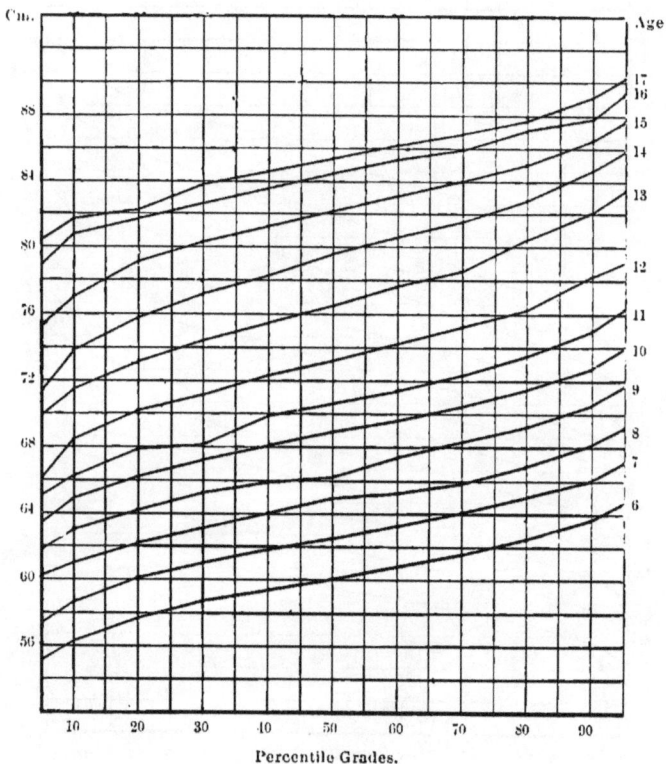

VOL. VI. NO. 12. PLATE X (from Table No. 19; pages 311, 314).
Girls' Height Sitting.

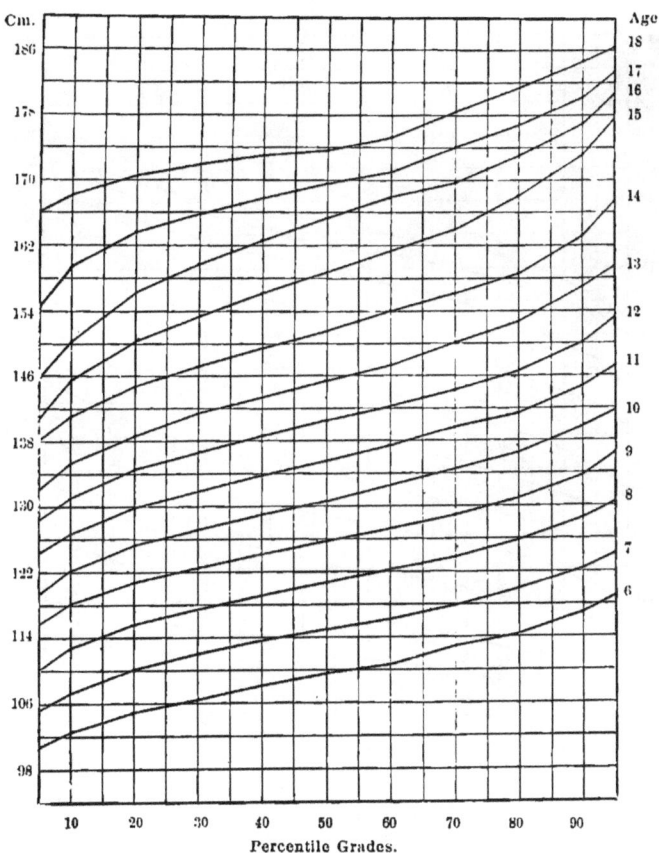

PLATE XI (from Table No. 20; pages 311, 315).
Boys' Span of Arms.

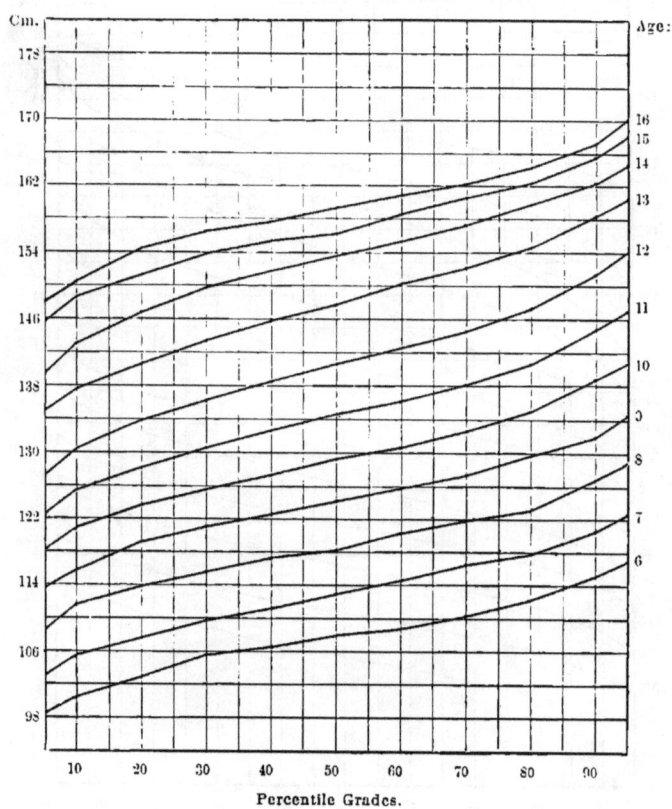

PLATE XII (from Table No. 20; pages 311, 315).
Girls' Span of Arms.

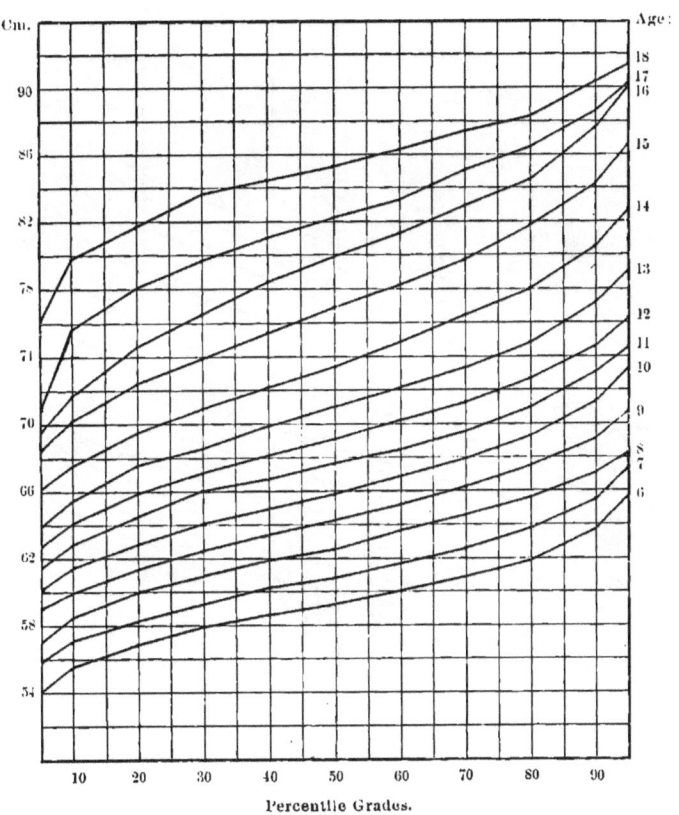

Boys' Girth of Chest.

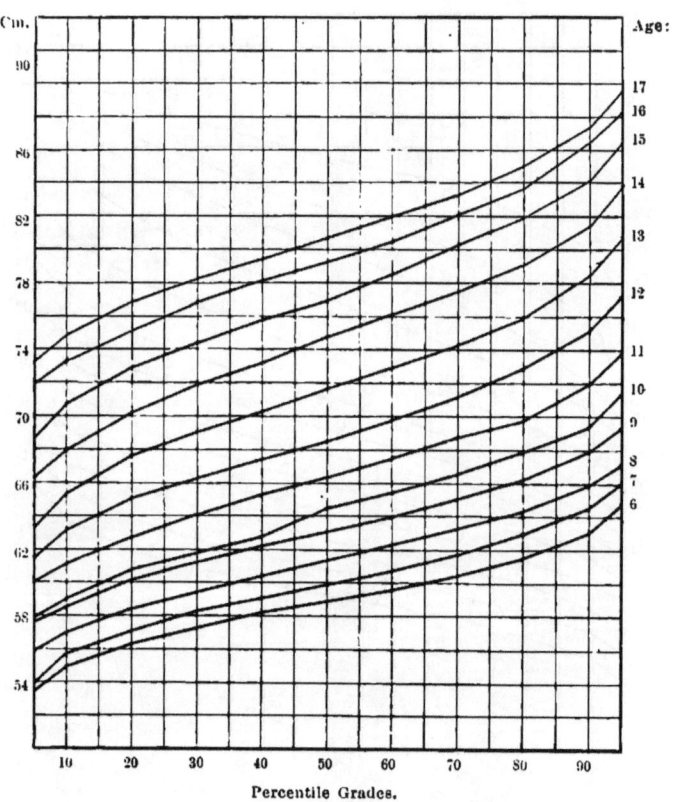

PLATE XIV (from Table No. 23; pages 311, 318).
Girls' Girth of Chest.

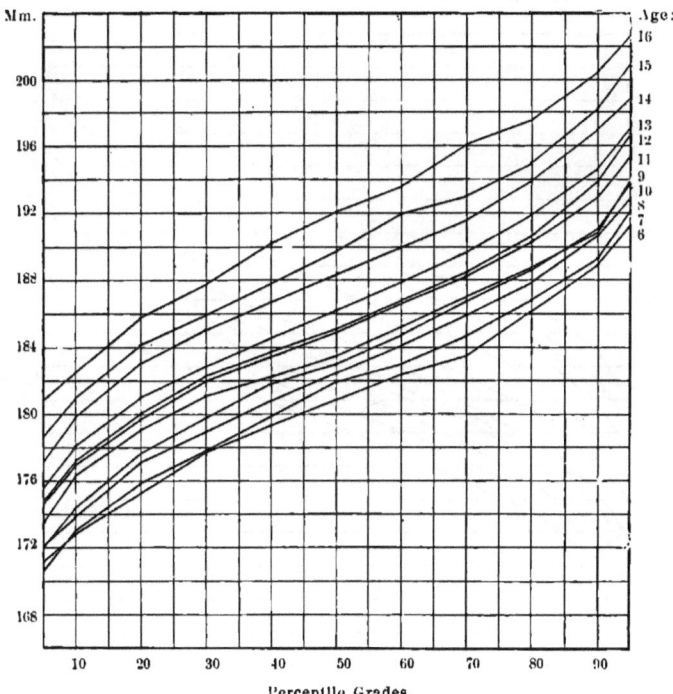

PLATE XV (from Table No. 24; pages 311, 319).
Boys' Length of Head.

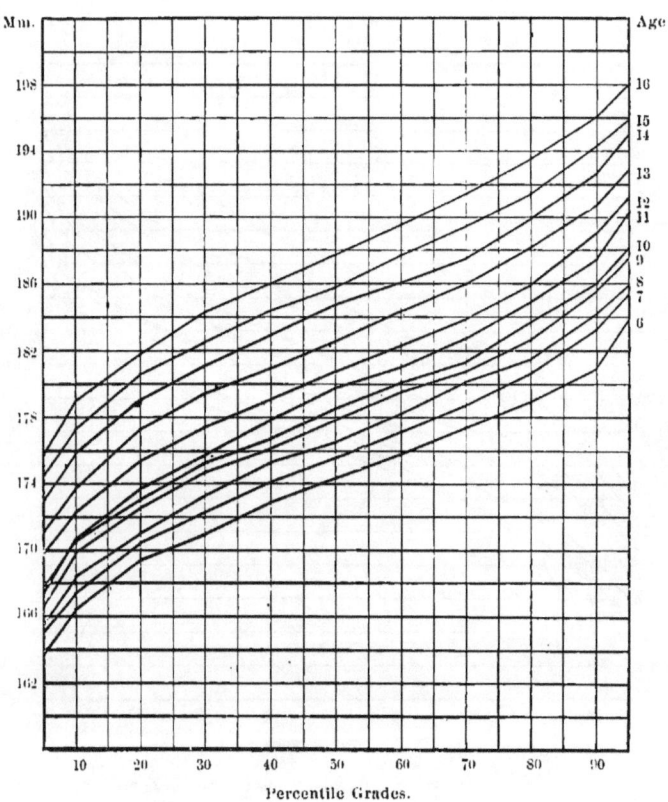

Vol. VI. No. 12. Plate XVI (from Table No. 24; pages 311, 319).
Girls' Length of Head.

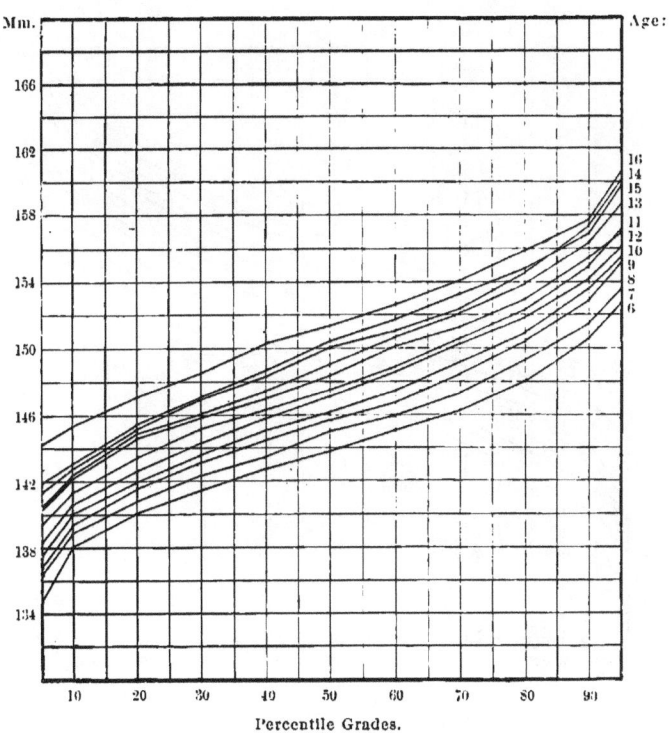

PLATE XVII (from Table No. 25; pages 311, 326).
Boys' Width of Head.

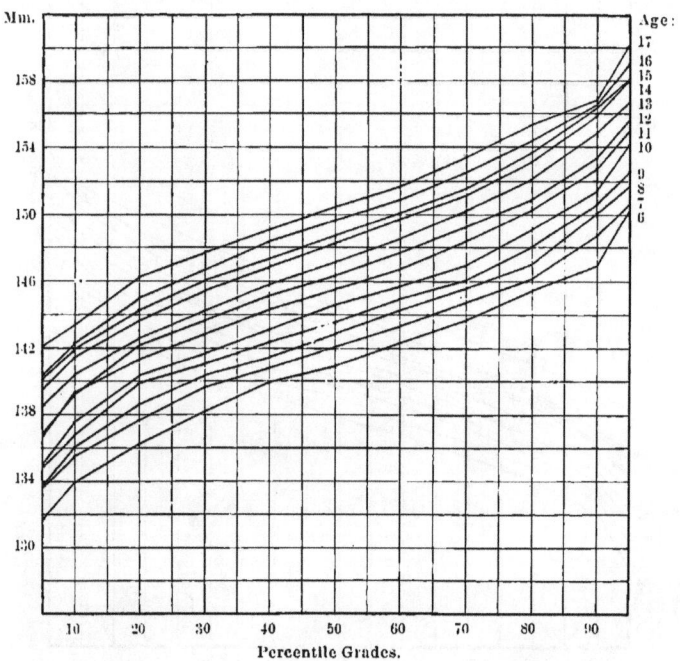

VOL. VI. No. 12. PLATE XVIII (from Table No. 25; pages 311, 320).
Girls' Width of Head.

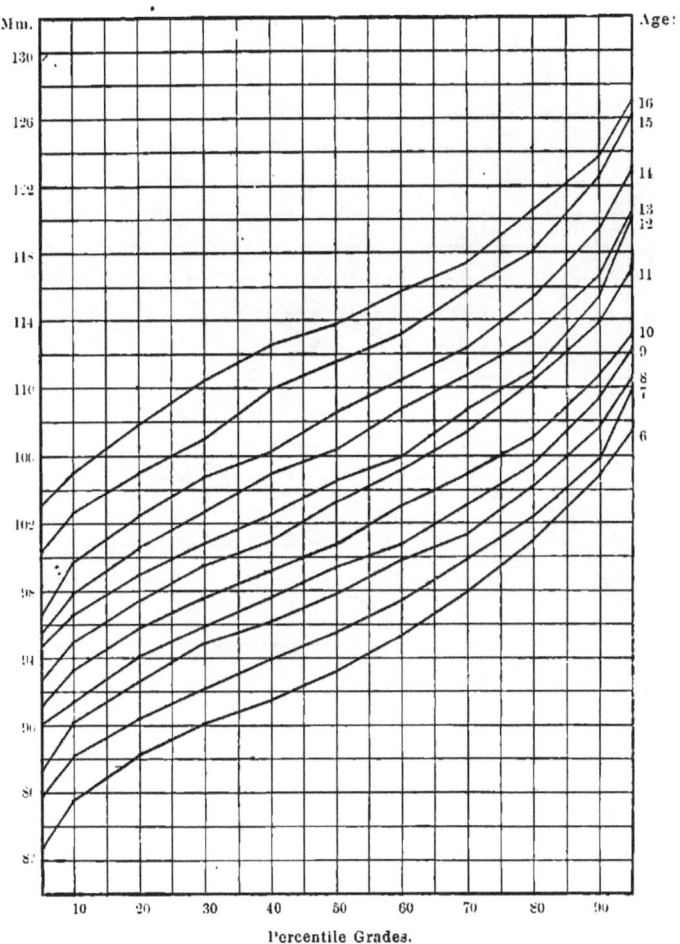

Boys' Height of Face from Root of Nose to Point of Chin.

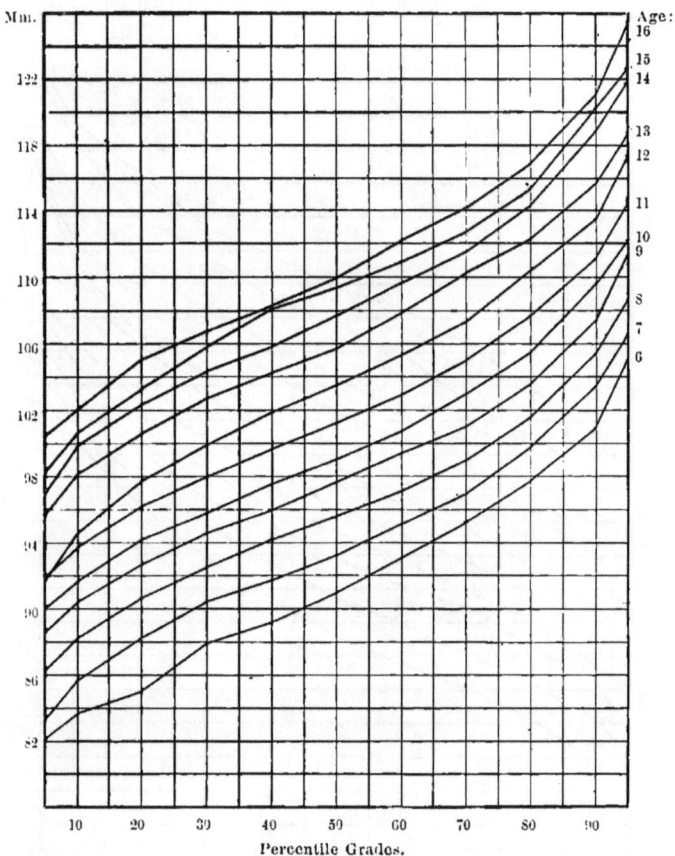

VOL. VI. NO. 12. PLATE XX (from Table No. 26; pages 311, 321).
Girls' Height of Face from Root of Nose to Point of Chin.

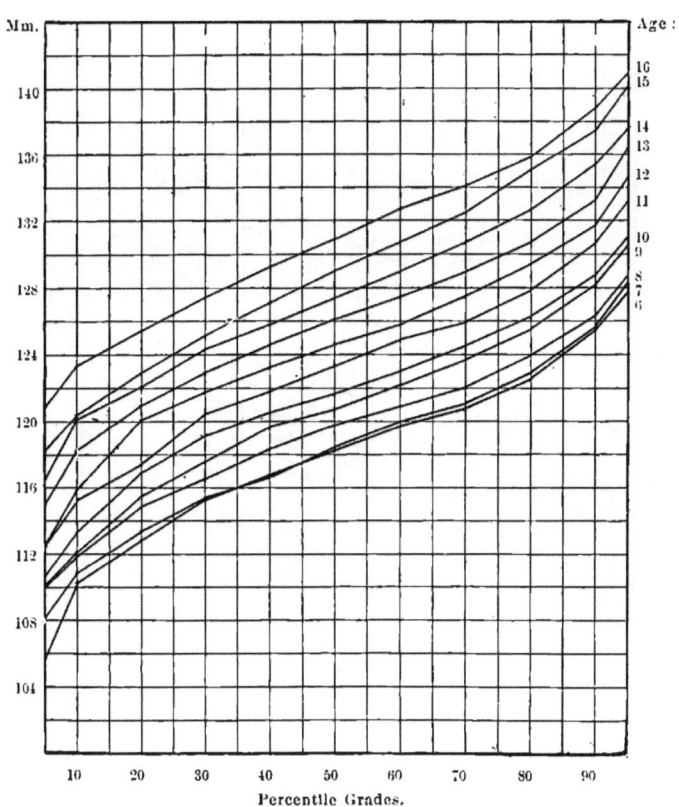

PLATE XXI (from Table No. 27; pages 311, 322).
Boys' Width of Face.

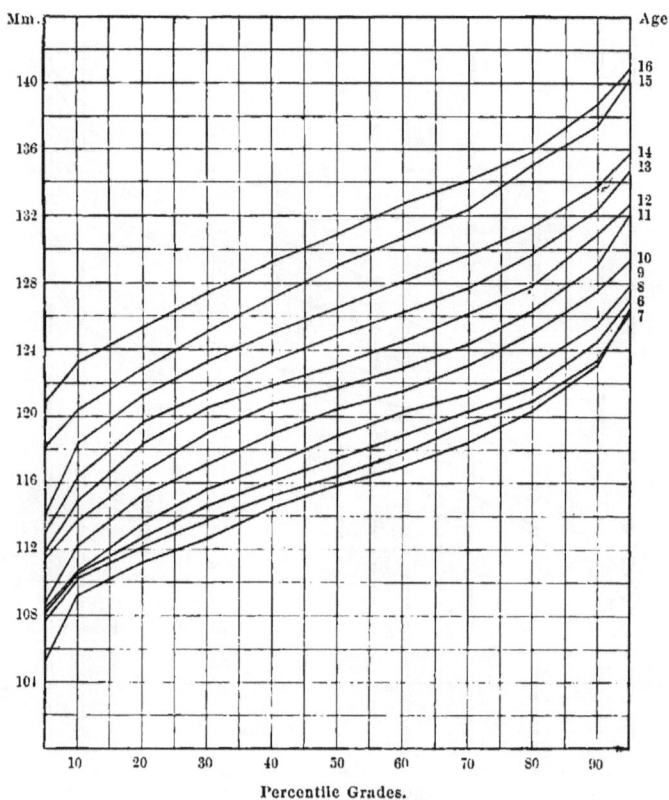

VOL. VI. NO. 12. PLATE XXII (from Table No. 27; page 322).
Girls' Width of Face.

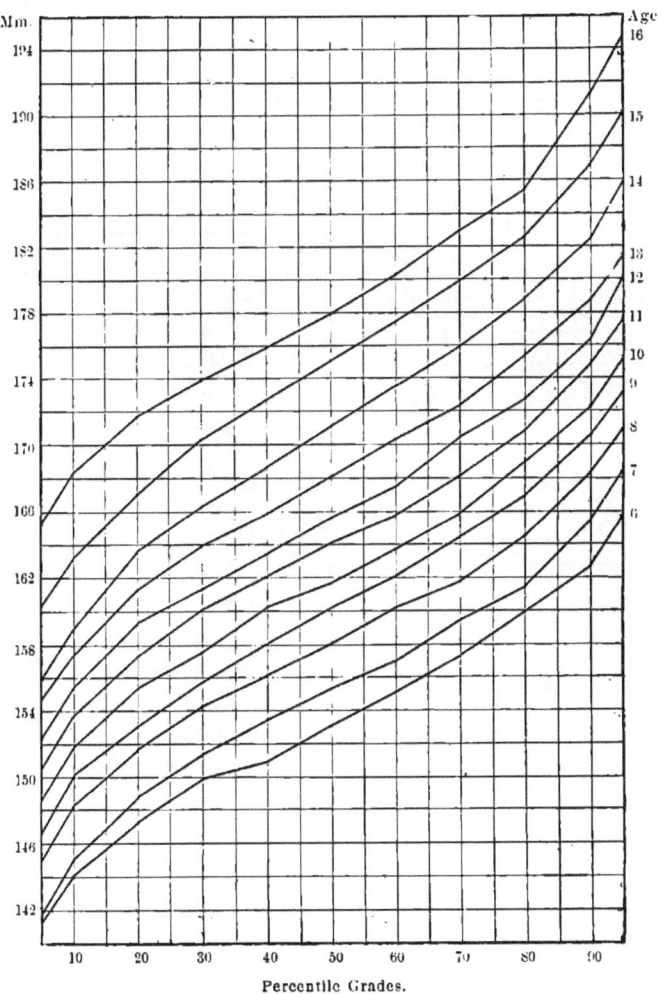

VOL. VI. NO. 12. Plate XXIII (from Table No. 28; pages 311, 323).
Boys' Height of Face from Hair-Line to Point of Chin.

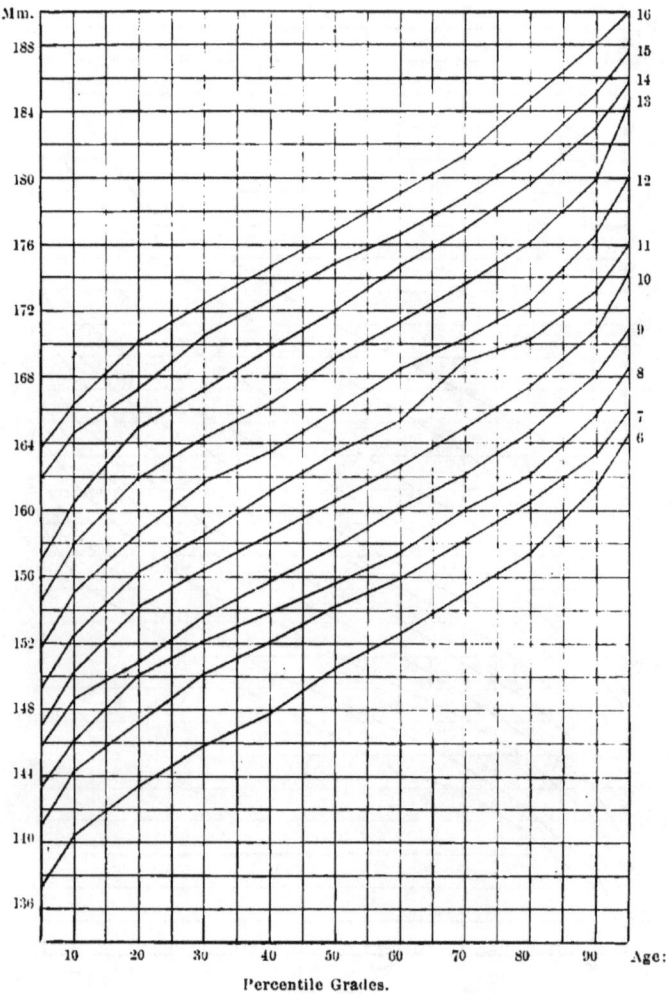

Vol. VI. No. 12. Plate XXIV (from Table No. 28; pages 311, 323).
Girls' Height of Face From Hair-Line to Point of Chin.

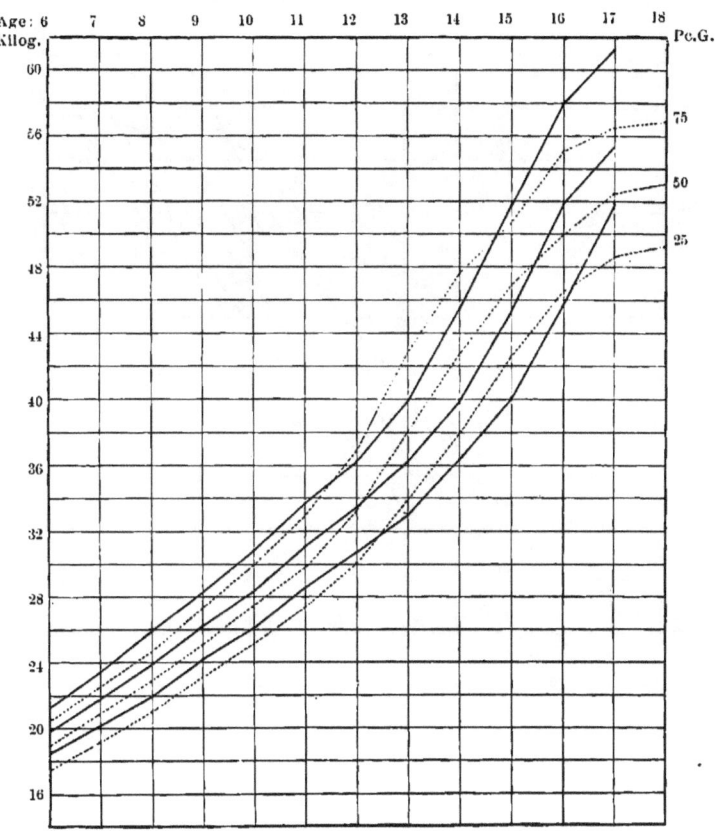

VOL. VI. NO. 12. PLATE XXV (from Table No. 39; pages 324, 351).
Weight.
75, 50 and 25 Percentile Grades.
Boys: Unbroken Lines. Girls: Broken Lines.

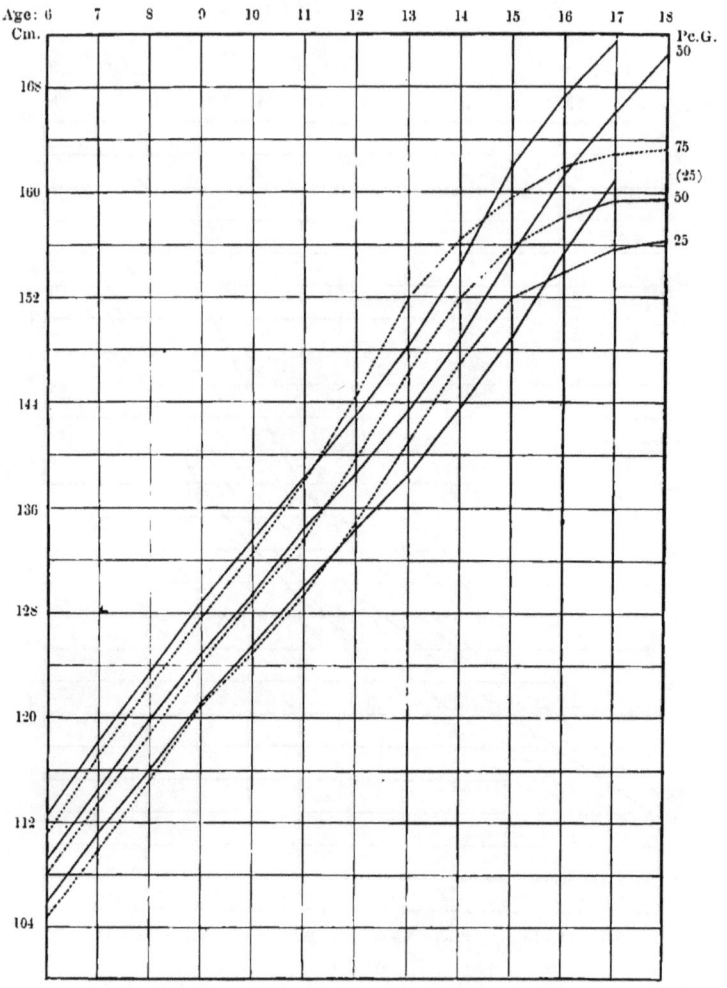

Vol. VI. No 12. Plate XXVI (from Table No. 38; pages 324, 350).
Height Standing.
75, 50 and 25 Percentile Grades.
Boys: Unbroken Lines.   Girls: Broken Lines.

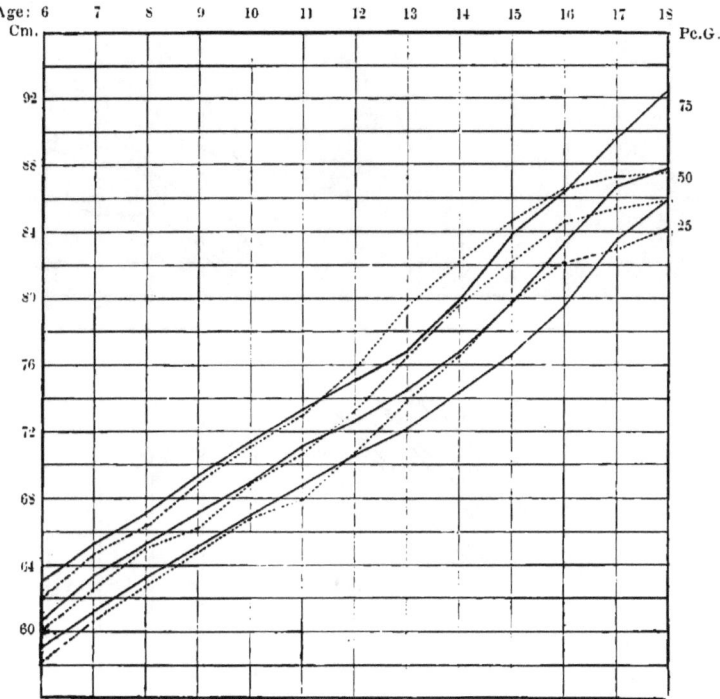

VOL. VI. NO. 12. PLATE XXVII (from Table No. 40; pages 324, 352).

Height Sitting.

75, 50 and 25 Percentile Grades.

Boys: Unbroken Lines.   Girls: Broken Lines.

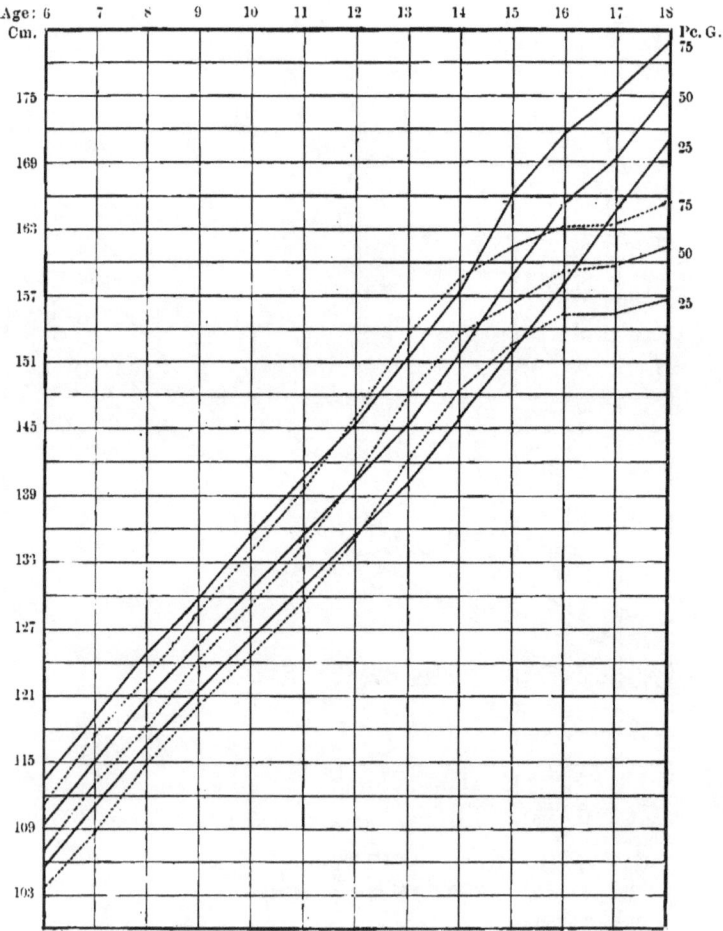

VOL. VI. NO. 12. PLATE XXVIII (from Table No. 41; pages 324, 353).
Span of Arms.
75, 50 and 25 Percentile Grades.
Boys: Unbroken Lines. Girls: Broken Lines.

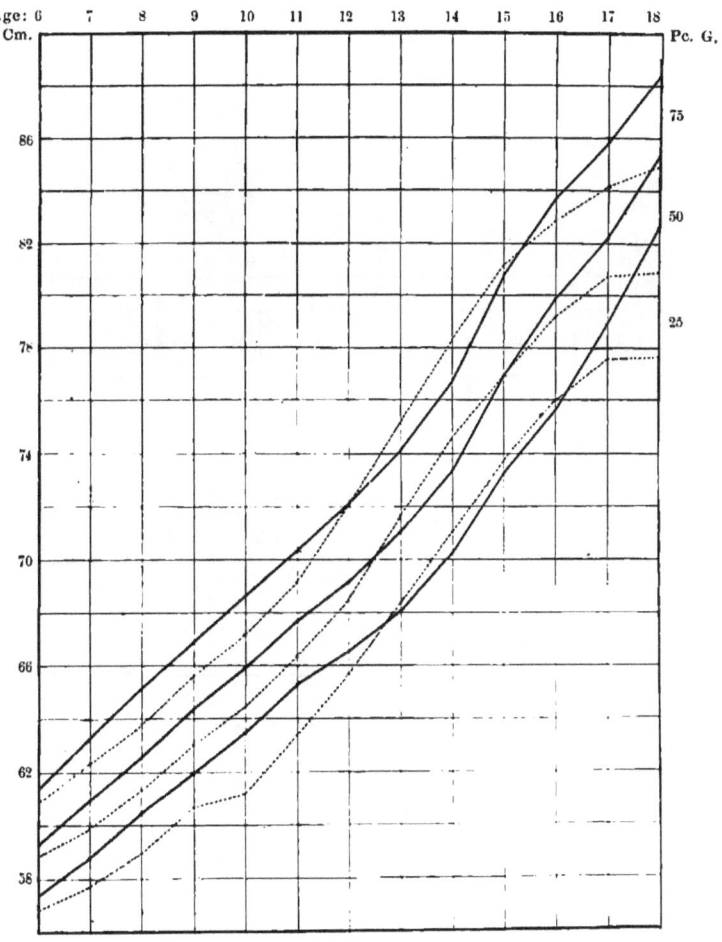

VOL. VI. NO. 12. PLATE XXIX (from Table No. 42; pages 324, 354).
Girth of Chest.
75, 50 and 25 Percentile Grades.
Boys: Unbroken Lines. Girls: Broken Lines.

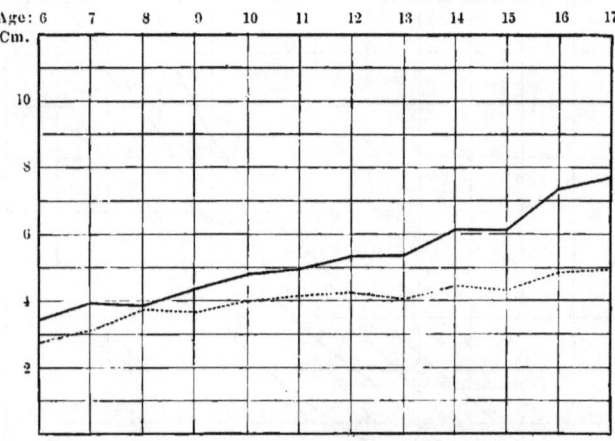

VOL. VI. NO. 12. PLATE XXX (from Tables No. 21, 22; pages 316, 317, 325).
Mean Expansion of Chest.
Boys: Unbroken Lines.   Girls: Broken Lines.

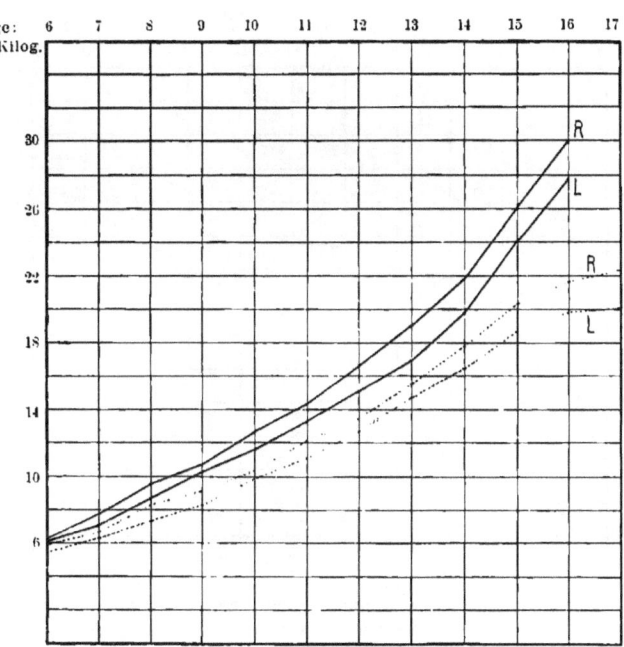

VOL. VI. NO. 12. PLATE XXXI (from Tables No. 43, 44; pages 825, 355, 356).
Mean Strength of Squeeze.

R: Right Hand.   L: Left Hand.
Boys: Unbroken Lines.   Girls: Broken Lines.

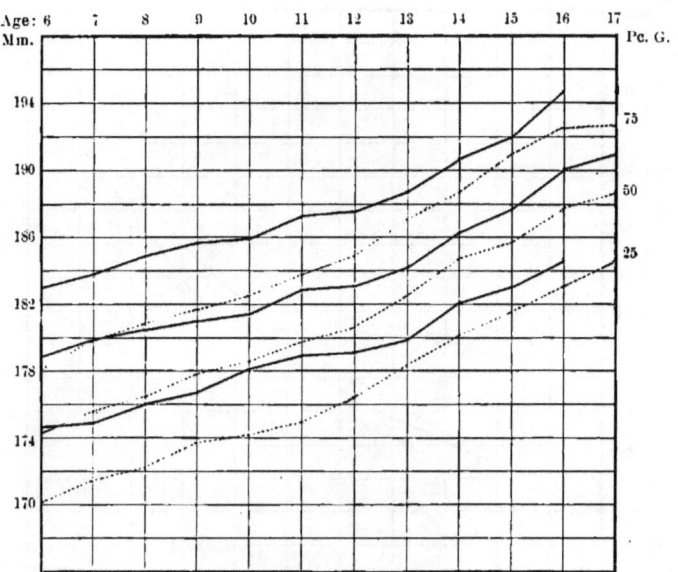

VOL. VI. NO. 12. PLATE XXXII (from Table No. 47; pages 325, 359).
Length of Head.
75, 50 and 25 Percentile Grades.
Boys: Unbroken Lines. Girls: Broken Lines.

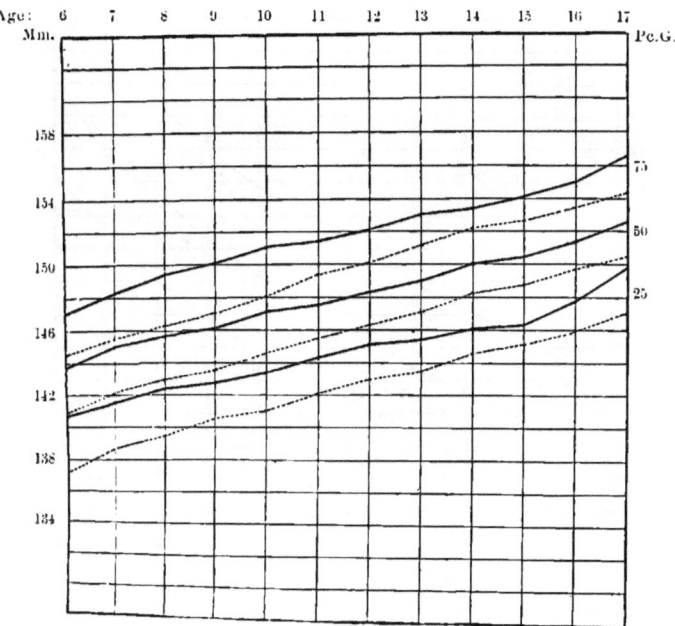

VOL. VI. NO. 12. PLATE XXXIII (from Table No. 18; pages 325, 360).
Width of Head.
75, 50 and 25 Percentile Grades.
Boys: Unbroken Lines. Girls: Broken Lines.

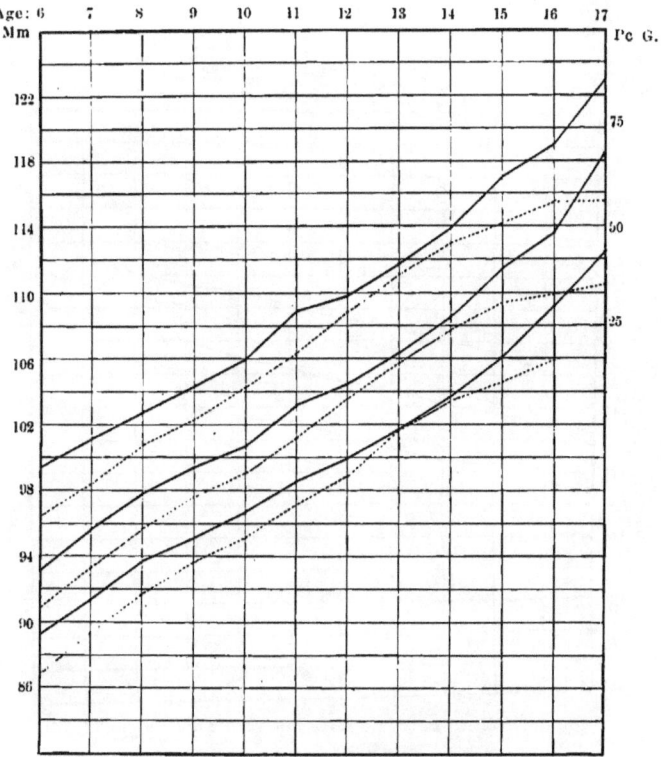

PLATE XXXIV (from Table No. 49; pages 325, 361).
Height of Face from Root of Nose to Point of Chin.
75, 50 and 25 Percentile Grades.
Boys: Unbroken Lines.   Girls: Broken Lines.

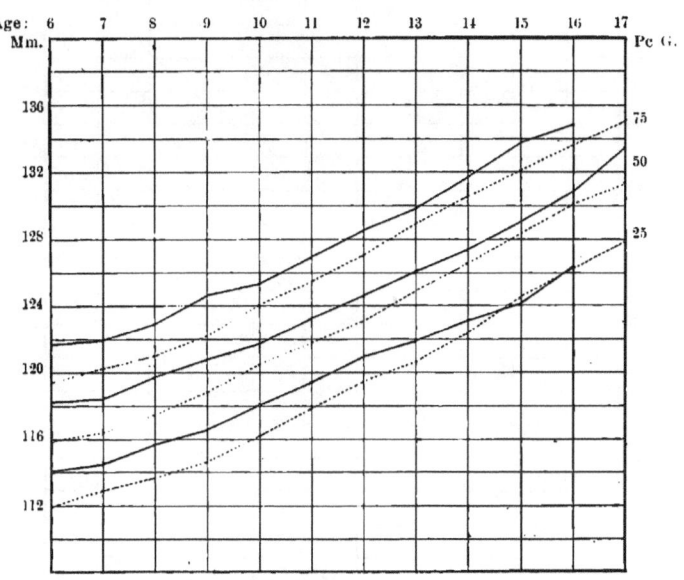

VOL. VI. NO. 12. PLATE XXXV (from Table No. 50; pages 325, 362).
Width of Face.
75, 50 and 25 Percentile Grades.
Boys: Unbroken Lines.   Girls: Broken Lines.

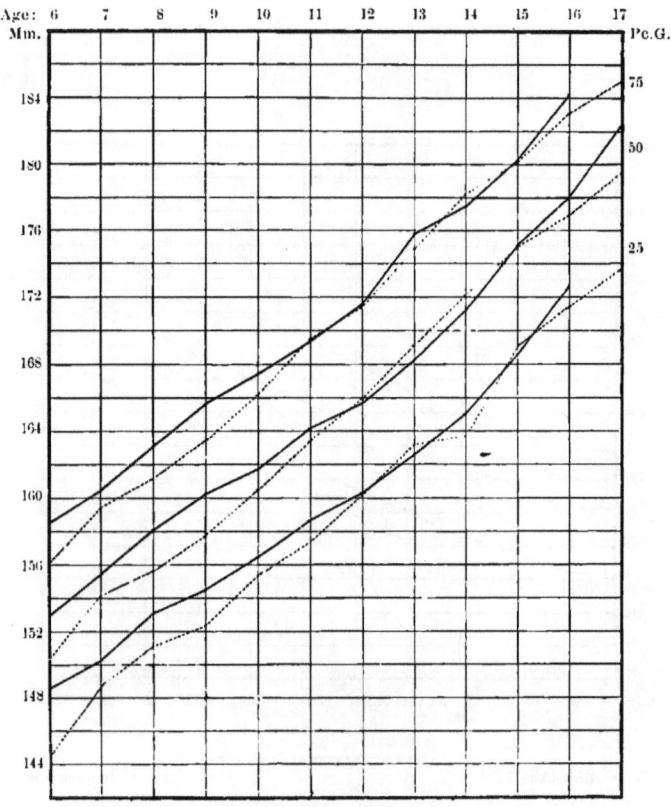

VOL. VI. NO. 12. PLATE XXXVI (from Table No. 51; pages 325, 363).
Height of Face from Hair-Line to Point of Chin.
75, 50 and 25 Percentile Grades.
Boys: Unbroken Lines.   Girls: Broken Lines.

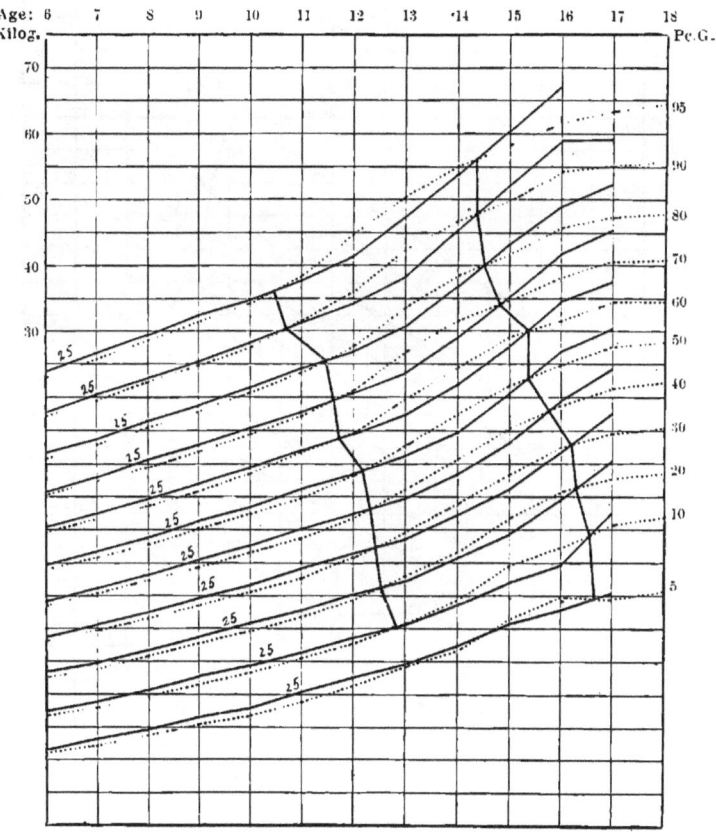

Plate XXXVII (from Table No. 17; pages 312, 325).
Percentile Grades.
Weight.
Boys: Unbroken Lines. Girls: Broken Lines.

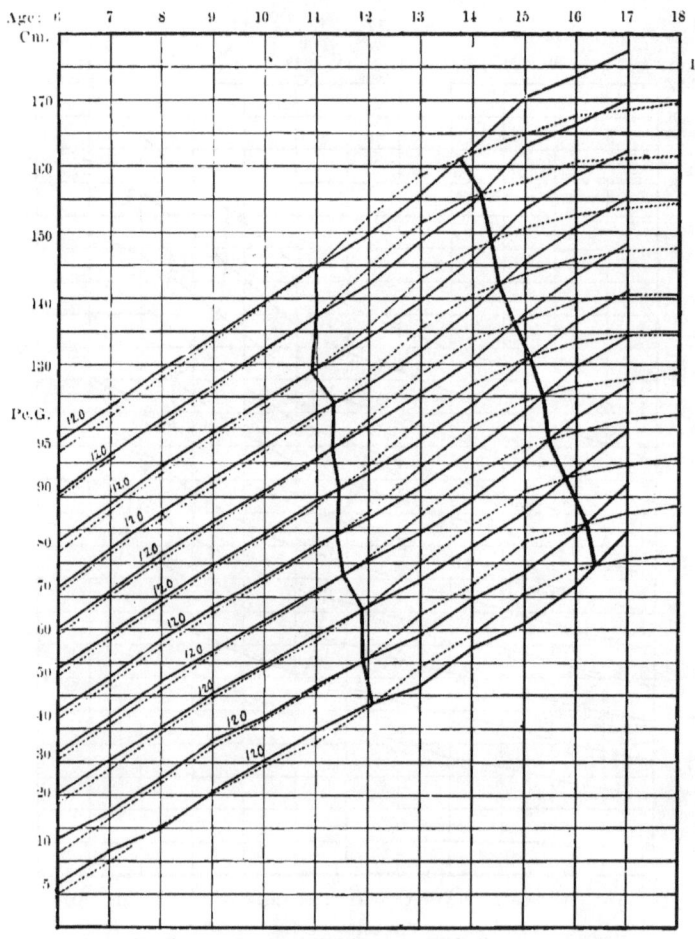

VOL. VI. NO. 12. PLATE XXXVIII (from Table No. 18; pages 313, 325).
Percentile Grades.
Height Standing.

Boys: Unbroken Lines.   Girls: Broken Lines.

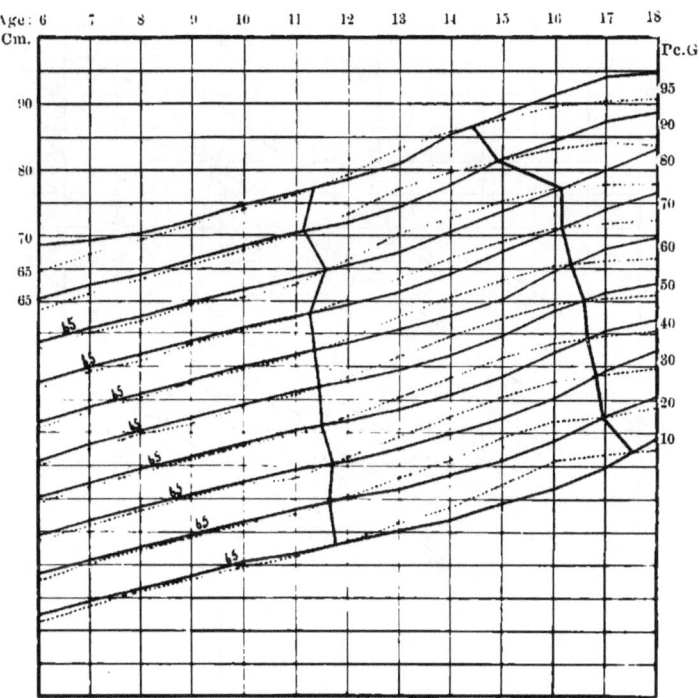

PLATE XXXIX (from Table No. 19; pages 314, 325).
Percentile Grades.
Height Sitting.
Boys: Unbroken Lines.   Girls: Broken Lines.

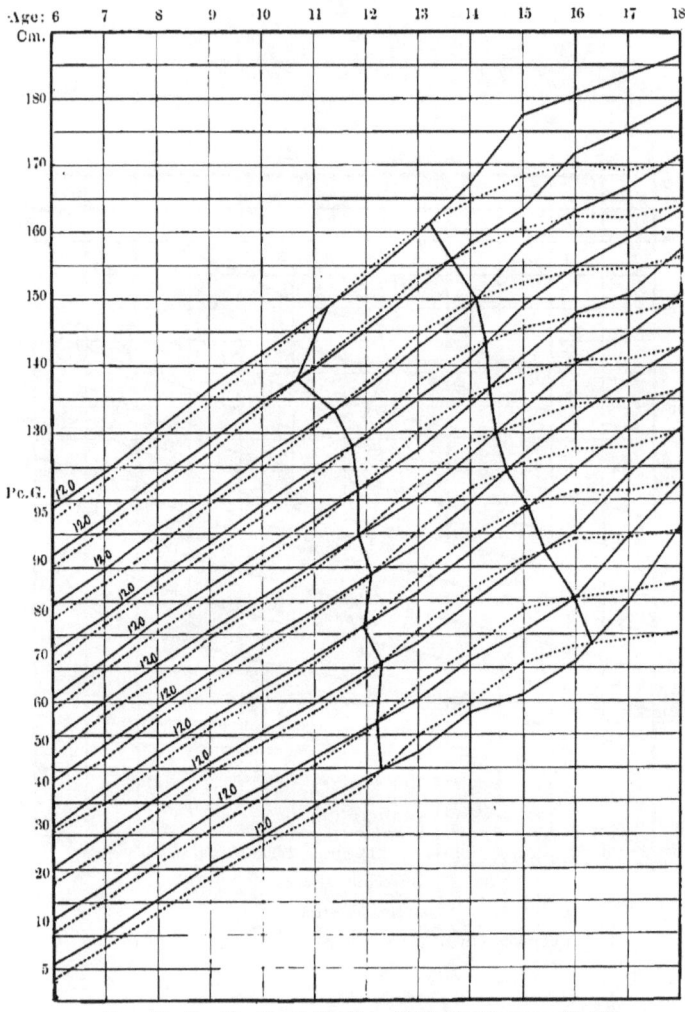

VOL. VI. NO. 12. PLATE XL (from Table No. 20; pages 315, 325).
Percentile Grades.
Span of Arms.
Boys: Unbroken Lines.   Girls: Broken Lines.

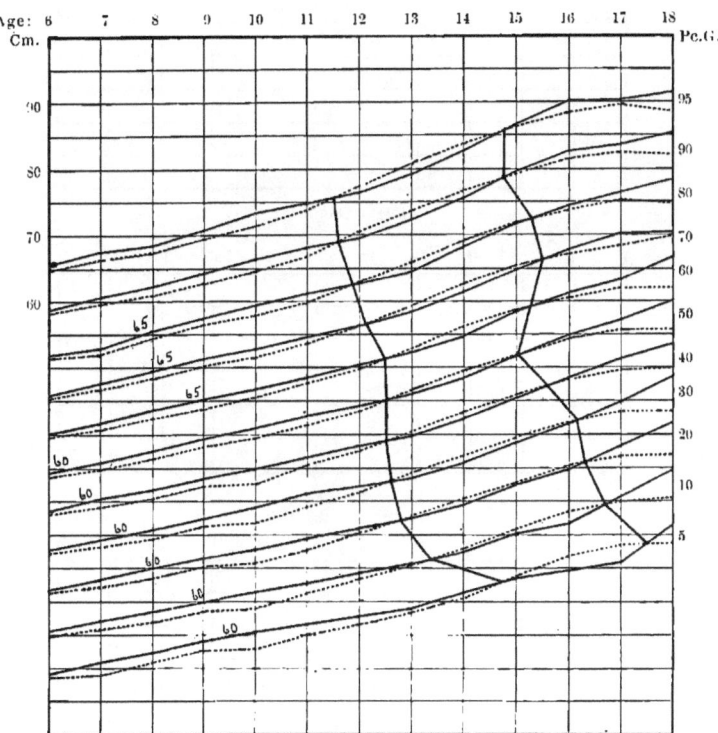

VOL. VI. NO. 12. PLATE XLI (from Table No. 23; pages 318, 325).
Percentile Grades.
Girth of Chest.

Boys: Unbroken Lines.   Girls: Broken Lines.

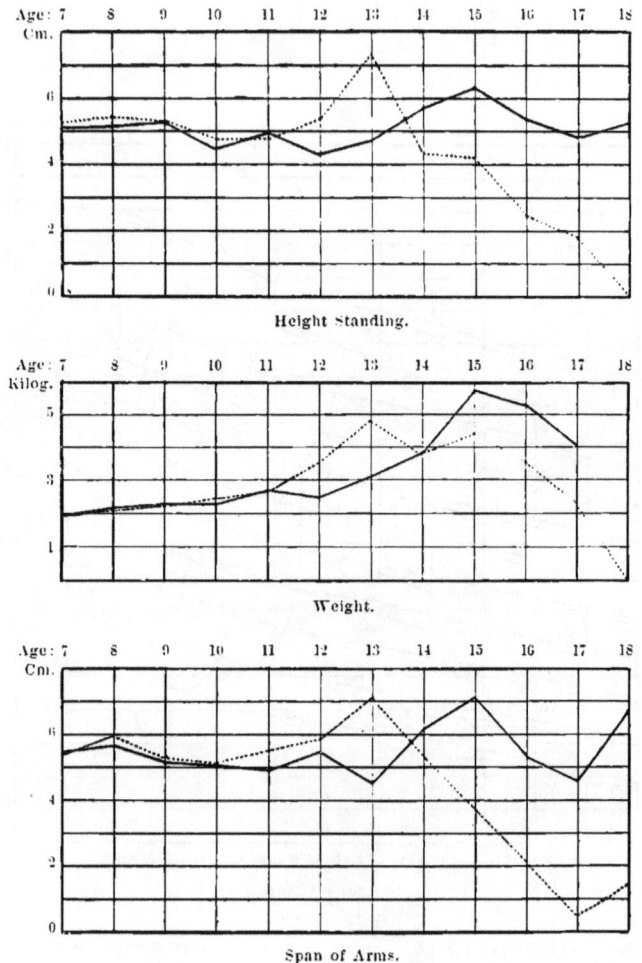

VOL. VI. NO. 12. PLATE XLII (from Tables No. 30, 31, 32; pages 327, 328, 329, 330.
Absolute Annual Increase.

Boys: Unbroken Lines.   Girls: Broken Lines.

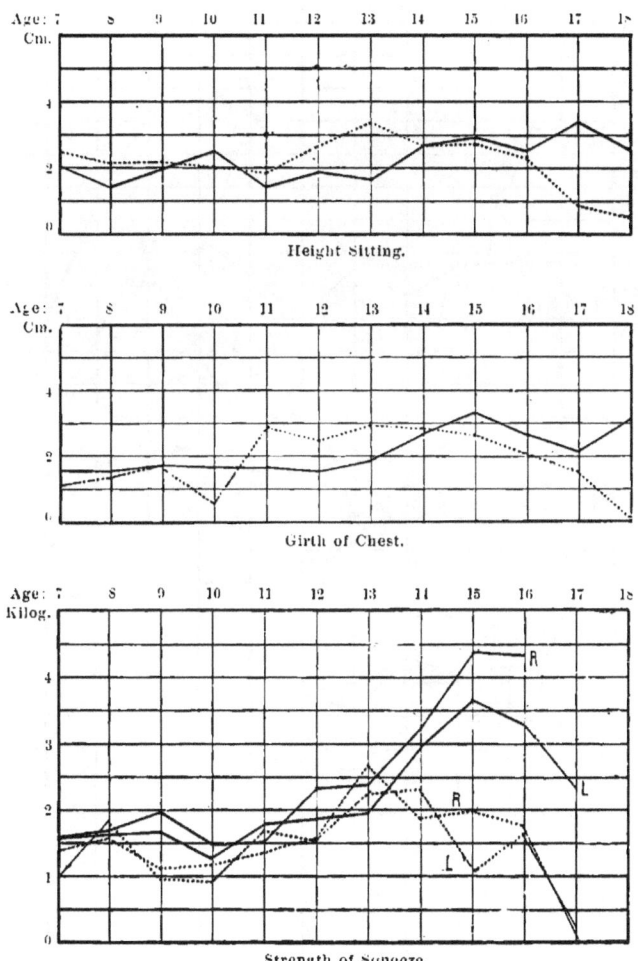

VOL. VI. NO. 12. PLATE XLIII (from Tables No. 33, 34, 43, 44, 45; pages 327, 331, 332, 355, 399)
Absolute Annual Increase.
Boys: Unbroken Lines.   Girls: Broken Lines.

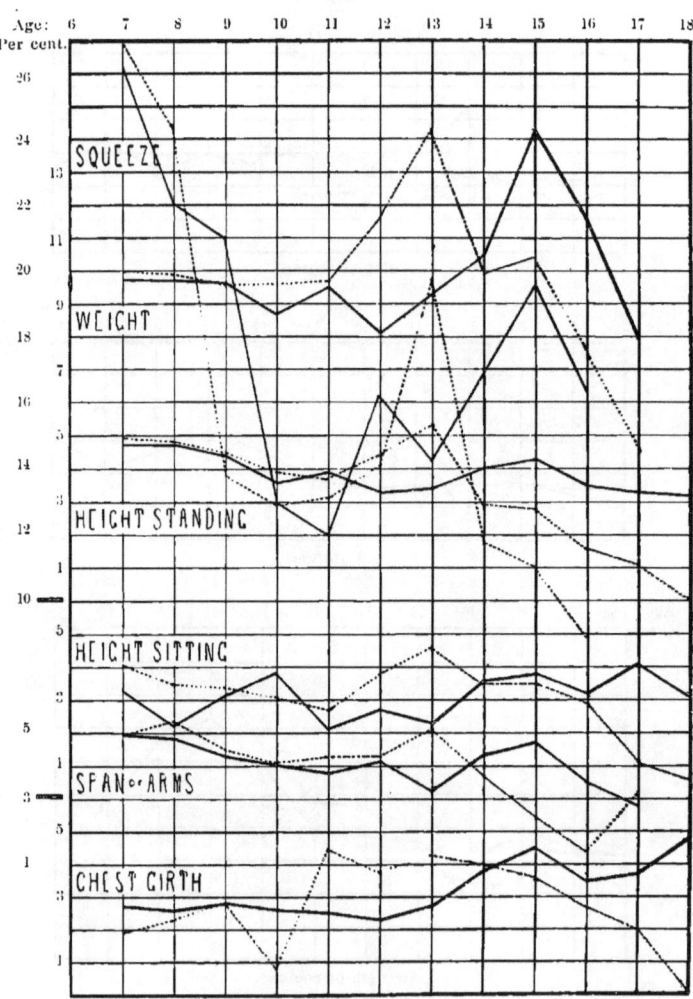

VOL. VI. NO. 12. PLATE XLIV (from Tables No. 38 to 43; pages 327, 333, 350 to 355).
Relative Annual Increase.
Boys: Unbroken Lines.   Girls: Broken Lines.

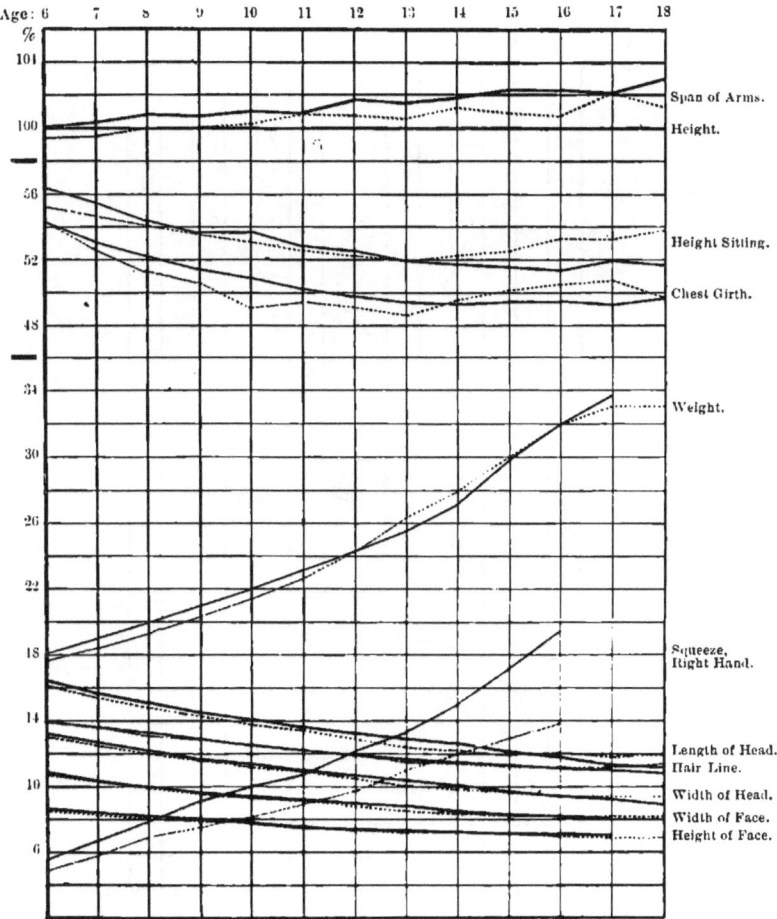

Vol. VI. No. 12. Plate XLV (from Table No. 38 to No. 51; pages 334, 350 to 363).
Relation of Average Weight, Span of Arms, Girth of Chest, etc., to Average Height.

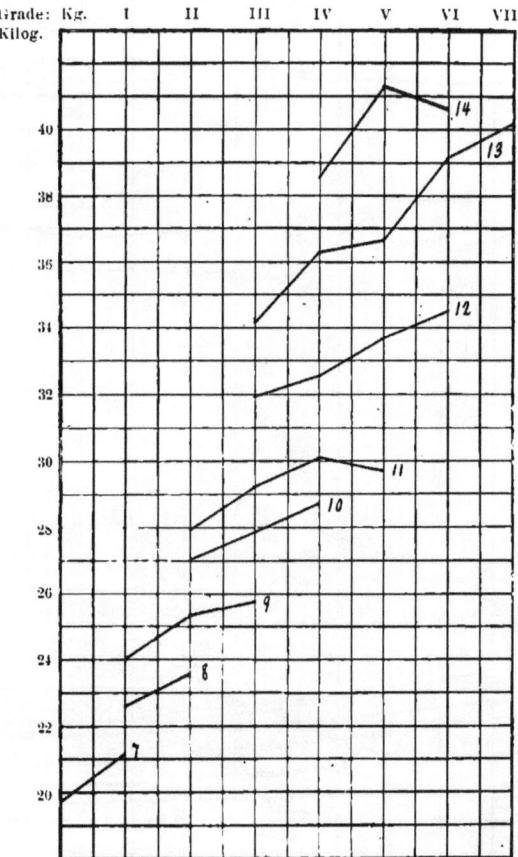

Vol. VI. No. 12. Plate XLVI (from Table No. 36; pages 336, 337).
The Weights of Daughters of Manual Tradesmen distributed by School Grade.

www.ingramcontent.com/pod-product-compliance
Lightning Source LLC
Chambersburg PA
CBHW031456160426
43195CB00010BB/993